INTRODUCTION
TO
PRACTICAL
MOLECULAR
BIOLOGY

INTRODUCTION
TO
PRACTICAL MOLECULAR BIOLOGY

Philippa D. Darbre

Imperial Cancer Research Fund, London, UK

A Wiley–Interscience Publication

Chichester   New York   Brisbane   Toronto   Singapore

Library of Congress Cataloging-in-Publication Data

Darbre, Philippa D.
  Introduction to practical molecular biology.

  'A Wiley—Interscience publication.'
  Includes index.
  1. Molecular biology — Technique.   I. Title.
QH506.D36   1988   574.8'8   88-10700
ISBN 0 471 91965 9 (pbk.)

British Library Cataloguing in Publication Data

Darbre, Philippa D.
  Introduction to practical molecular biology
  1. Molecular biology
  I. Title
  574.8'8

  ISBN 0 471 91965 9

Typeset by MHL Typesetting Ltd, Coventry
Printed and bound in Great Britain

Contents

Preface

Molecular biology has become a very powerful tool in many fields of biological, biochemical and medical science. It is now used very widely both in fundamental research and for diagnosis of disease. This methodology has enabled major advances to be made in medical research resulting in the better diagnosis and understanding of blood disorders, immunological deficiencies, cystic fibrosis and cancer. It has helped research into animal diseases and there is now also wide application in botanical fields. More and more scientists are discovering a need to use these techniques. Yet for many people who are not established molecular biologists, it is an awesome task to know where to start. It is for just such people that this book has been written.

This book was developed from a collection of laboratory protocols, which have been used successfully in the laboratory over the past five years. Simple protocols for the preparation of DNA and RNA are described, followed by a step-by-step procedure for DNA analysis by Southern blotting and RNA analysis by Northern blotting. Each chapter begins with a simple explanation of the principles involved. This is followed by a plan of how to fit the successive stages into a weekly schedule, a list of equipment and materials required, and finally a simple-to-follow protocol. Here lies the way for scientists wishing to take up molecular biology.

This volume is not intended to provide a major laboratory manual for the established and highly experienced molecular biologist but is an introduction to the field. It concentrates on simple protocols giving advice on the problems which might be encountered. It is assumed that genes to be used have already been cloned. Procedures are described for the amplification of cloned DNA and the preparation of probes required. No attempt is made to discuss the more complex strategies for gene cloning or engineering. For excellent comprehensive manuals, where answers to more detailed questions will be found, the reader is referred to the books by Maniatis *et al.** and Ausubel *et al.***

The production of reliable protocols for experiments in any scientific discipline poses the problem of acknowledgements. All relevant references quoted in the text are to original procedures but it is inevitable that methods are continually being modified and improved. In this latter respect, I am indebted to my many colleagues at the laboratories of the Imperial Cancer Research Fund, London, for their advice and support.

Finally, I want to express by deep gratitude to my family who gave encouragement when it was most needed − to my husband, my daughter and my parents without whose support this book might never have been completed.

Philippa D. Darbre

  * Maniatis, T., Fritsch, E.F. and Sambrook, J. (1982) *Molecular Cloning: A Laboratory Manual*, Cold Spring Harbor Laboratory, Cold Spring Harbor, NY, USA.
** Ausubel, F.M., Brent, R., Kingston, R.E., Moore, D.D., Seidman, J.G., Smith, J.A. and Struhl, K. (1987) *Current Protocols in Molecular Biology*, Green Publishing Associates and Wiley−Interscience, New York, USA.

Chapter 1

Introduction

Twenty years ago, DNA was one of the most difficult biochemical macromolecules to analyse and it appeared to be a daunting task to try to understand the function of such an enormously long nucleotide structure. However, today the situation has entirely changed and DNA has become the easiest macromolecule of the cell to study. This rapid development of modern molecular biology and the so-called recombinant DNA technology has resulted from a unique combination of new discoveries (new enzymes and nucleic acid hybridisation) with old techniques (microbial genetics) and these will be described in simple terms in this book.

DNA can be easily purified from a cell and once isolated it is much more stable than many other macromolecules. It can be cut very precisely and reproducibly with restriction enzymes, enabling excision of specific pieces of DNA, which can then be obtained in essentially unlimited quantities (DNA cloning). Hybridisation techniques and rapid sequencing methods for cloned DNA have enabled determination of the structure and organisation of large parts of the genome of many organisms. It is now within the realms of possibility to obtain a sequence for the entire structure of the human genome.

RNA can also be easily purified, although being less stable than DNA, it has to be handled with greater care. With the use of cloned DNA, both qualitative and quantitative studies of any specific RNA can be made. This approach has provided a powerful tool in the study of eukaryotic gene expression.

The early years following production of the first recombinant DNA molecules did not open up a whole new era of research but rather were filled with fears about the safety of such powerful technology. At the Asilomar conference in 1975, the landmark guidelines for the use of recombinant DNA were finally drawn up. Fortunately today most forms of such research are no longer subject to any preventative form of regulation.

NUCLEIC ACID STRUCTURE

A general introduction to molecular biology must inevitably begin with a brief outline of nucleic acid structure. Both deoxyribonucleic acid (DNA) and ribonucleic acid (RNA) are chainlike macromolecules made up of strings of monomeric units called nucleotides. The monomeric units of DNA are called deoxyribonucleotides, those of RNA are ribonucleotides. Each nucleotide is composed of three components: (1) a nitrogenous base, which is either a pyrimidine or purine derivative; (2) a pentose sugar; and (3) a molecule of phosphoric acid. DNA and RNA differ both in the sugar present and in their bases. The sugar 2-deoxy-D-ribose is present in DNA, whereas D-ribose is present in RNA. The four bases present in DNA are adenine and guanine (purines), cytosine and thymine (pyrimidines). RNA has the bases adenine, guanine and cytosine but uracil instead of thymine. The structure of these five bases is given in Fig. 1.1 and the general structure of the nucleotides in Fig. 1.2.

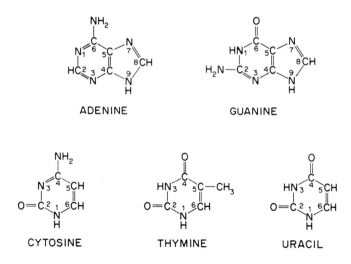

Fig. 1.1 Bases present in nucleic acids. There are two purines — adenine and guanine, and three pyrimidines — cytosine, thymine and uracil

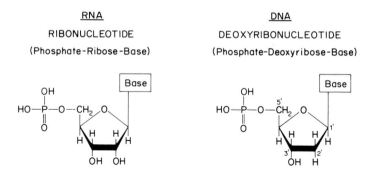

Fig. 1.2 The general structure of nucleotides in DNA and RNA

DNA consists, then, of covalently linked chains of deoxyribonucleotides and RNA of chains of ribonucleotides. The linkage is by phosphodiester bridges between the 5′-hydroxyl group of the phosphate on one nucleotide and the 3′-hydroxyl group of the sugar molecule of the next nucleotide. The backbone of both DNA and RNA is thus composed of alternating phosphate and pentose groups, with the bases acting as distinct side-chains.

In 1953, Watson and Crick made the landmark discovery of the double-helical structure of DNA. In this structure, two polynucleotide chains are wound into a helix such
that the backbone strands are of alternating phosphate and sugar residues, leaving
the bases projecting perpendicularly into the centre axis. The bases of the two chains
are then held together very specifically by hydrogen bonding such that adenine can
only pair with thymine, and cytosine only with guanine. This structure was based on
X-ray data from Franklin and Wilkins, together with Chargaff's observations that in
DNA the ratio of adenine to thymine and of cytosine to guanine was always very close
to 1.0.

Thus, in DNA there are two chains coiled around a common axis and held together
by base pairing between adenine and thymine, and between cytosine and guanine.
This produces two chains which are not identical but, because of base pairing, are
precise complements of each other. In addition, the chains do not run in the same
direction with respect to their internucleotide linkages but rather are antiparallel. That
is, if two adjacent deoxyribonucleotides T and C in the same chain are linked 5'-3',
the complementary deoxyribonucleotides A and G in the other chain will be linked
3'-5'. This structure is outlined in Fig. 1.3.

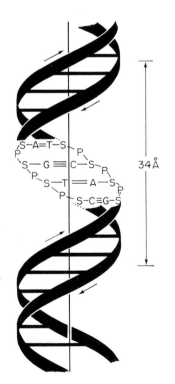

Fig. 1.3 Double-helical structure of DNA. S = sugar; P = phosphate; A = adenine; C = cytosine;
G = guanine; T = thymine

For RNA, there are three major types in cells: messenger RNA (mRNA); transfer RNA (tRNA); and ribosomal RNA (rRNA). The most abundant species is rRNA, making up about 80% of the total; next is tRNA, making up 15%; mRNA is present in the smallest amounts — 5% or less. Messenger RNA is synthesised in the nucleus during the process of transcription, in which the sequence of bases in one strand of DNA is enzymatically transcribed into the form of a single strand of mRNA with complementary base sequence. mRNA is thus composed of only the four bases — adenine, cytosine, guanine and uracil. After transcription, the RNA is processed and passed out into the cytoplasm, where it then serves at the ribosomes as the template for sequential ordering of amino acids during protein synthesis. Although mRNA molecules make up only a very small fraction of the total RNA of the cell, there is a wide variety of them, differing greatly in molecular weight and in base sequence.

tRNAs are relatively small molecules that act as the carriers of specific amino acids during protein synthesis. They have similar molecular weights in the range of 23 000 to 30 000 (sedimentation coefficient of about 4S). These RNAs characteristically contain a rather large number of minor or 'odd' bases, and there are one or more tRNAs for each amino acid found in proteins.

rRNA is the most abundant RNA and is found in the ribosomes of eukaryotic cells in three forms, with sedimentation coefficients of 28S, 18S and 5S.

RESTRICTION ENZYMES

Many bacteria make enzymes called restriction endonucleases, which protect them by degrading any invading foreign DNA molecules. Each enzyme recognises a specific sequence within double-stranded DNA, which is typically four to six nucleotides in length with a two-fold axis of symmetry. The corresponding sequences in the genome of the bacterium itself are protected by methylation at an adenine or cytosine residue, but any foreign DNA molecule is immediately recognised and cut to pieces.
Many restriction endonucleases have now been purified from prokaryotes and are commercially available. The very specific cleavage of DNA by these enzymes and the availability of such a wide variety of enzymes cutting at different nucleotide sequences have been very important in the development of molecular biology.

Each restriction enzyme will cut any length of DNA double helix into a series of fragments, known as restriction fragments. However, many of these enzymes do not cut DNA exactly at the axis of dyad symmetry and thus the cutting can be broadly divided into three types:

1. Enzymes which cut DNA in a staggered fashion producing restriction fragments with protruding cohesive 5′ termini.

$$\text{---G}^{\downarrow}\text{A-A-T-T-C---}\xrightarrow{\text{Eco R1}}\text{---G} + \text{A-A-T-T-C---}$$

5' ---G A-A-T-T-C--- 3'
3' C-T-T-A-A G--- 5' →(Eco R1) 5' ---G 3' + 5' A-A-T-T-C--- 3'
3' C-T-T-A-A 5'   3' G 5'

2. Enzymes which cut DNA in a staggered fashion but in the opposite way to produce restriction fragments with protruding cohesive 3′ termini.

5' ---C-T-C-G-A G--- 3'
3' ---G A-G-C-T-C--- 5' →(Pst I) 5' ---C-T-C-G-A 3' + 5' G--- 3'
3' ---G 5'   3' A-G-C-T-C--- 5'

3. Lastly, there are some enzymes which cleave exactly at the axis of dyad symmetry to produce blunt-ended fragments.

5' ---T-G-G C-C-A--- 3'
3' ---A-C-C G-G-T--- 5' →(Bal I) 5' ---T-G-G 3' + C-C-A--- 3'
3' ---A-C-C   G-G-T--- 5'

In general, different restriction enzymes recognise different sequences, but there are some examples of enzymes isolated from different sources that cleave within the same target sequences. These are known as isoschizomers.

Each enzyme has a set of optimal reaction conditions, nowadays given on the information sheet supplied by the manufacturer. The major variables are the temperature and salt composition, although some enzymes are affected by methylation of the DNA, in particular of cytosine residues.

OTHER ENZYMES

Many enzymes, apart from restriction endonucleases, are used widely in molecular biology and the actions of those referred to in this book are summarised below.

Ribonucleases A and T$_1$

(Used in preparation of DNA, Chapter 2.)
Ribonuclease A (bovine pancreas) is an endoribonuclease which attacks pyrimidine nucleotides at the 3'-phosphate group and cleaves the 5'-phosphate linkage to the adjacent nucleotide.

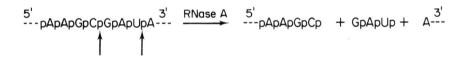

Ribonuclease T$_1$ (*Aspergillus oryzae*) is an endoribonuclease which attacks RNA in a similar fashion to ribonuclease A but at the 3'-phosphate group of guanine-containing nucleotides.

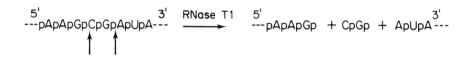

Deoxyribonuclease I (DNase I)

(Used in nick translation of DNA, Chapters 3 and 5.)
This enzyme is an endonuclease that hydrolyses double- or single-stranded DNA to a complex mixture of polynucleotides with 5'-phosphate termini. In the presence of magnesium ions, it attacks each strand of DNA independently giving a random distribution of cleavage.

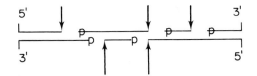

E. coli DNA polymerase I

(Used in nick translation of DNA, Chapters 3 and 5.)
This enzyme carries three separate enzyme activities. It carries a 5' → 3' polymerase activity, adding nucleotide residues to the 3'-hydroxyl terminus that is created when one strand of a double-stranded DNA molecule is damaged, for example:

	+ d ATP	
5' OH	+ d CTP	5'
---C-C-G	+ d GTP	---C-C-G-A-T-C-G---
---G-G-C-T-A-G-C---	+ d TTP	---G-G-C-T-A-G-C---
3'	+ Mg^{2+}	3'

However, it also carries a $5' \rightarrow 3'$ exonuclease activity and a $3' \rightarrow 5'$ exonuclease activity. This makes it extremely useful in the nick translation reaction since it can both remove nucleotides from the 5' side of the nick and add them to the 3' side. By replacing the pre-existing nucleotides with radiolabelled (^{32}P) nucleotides, it is possible to prepare ^{32}P-labelled DNA of specific activities of 10^8 cpm/μg.

Alkaline phosphatase

(Used in labelling DNA molecular weight markers, Chapter 3.)
This enzyme can be prepared from either bacteria or calf intestine; it catalyses the removal of 5'-phosphate residues from fragments of DNA or RNA of any length, either single- or double-stranded.

$$5'_p\ \text{DNA} \xrightarrow[\text{phosphatase}]{\text{Alkaline}} 5'_{OH}\ \text{DNA} + p$$

T4 polynucleotide kinase

(Used in labelling DNA molecular weight markers, Chapter 3.)
This enzyme is prepared from T4-infected *E. coli* and catalyses the transfer of the γ-phosphate of ATP to a free 5'-hydroxyl terminus in DNA or RNA, either single- or double-stranded. Since it reverses the action of alkaline phosphatase, it is possible to radiolabel any piece of nucleic acid by removing phosphate with alkaline phosphatase and adding radiolabelled phosphate with polynucleotide kinase.

$$5'_{OH}\ \text{DNA} + \gamma\text{-}^{32}\text{P-ATP} \xrightarrow[\text{+Mg}^{2+}]{\substack{\text{Polynucleotide} \\ \text{kinase}}} 5'_{32p}\ \text{DNA} + \text{ADP}$$

T4 DNA ligase

(Used in joining DNA molecules while cloning, Chapter 6.)
This enzyme is prepared from T4-infected *E. coli* and catalyses the formation of a phosphodiester bond between adjacent 3'-OH and 5'-P termini in double-stranded DNA. It can thus be used for joining together DNA molecules that are either blunt-ended or have cohesive termini sufficiently compatible to allow base pairing to one another.

$$
\begin{array}{l}
5'\text{---T-A-G}^{OH} \\
3'\text{---A-T-C}_{p}
\end{array}
+
\begin{array}{l}
{}^{p}\text{C-G-A---}3' \\
\text{G-C-T---}5' \\
{}_{OH}
\end{array}
\xrightarrow{\text{DNA ligase}}
\begin{array}{l}
5'\text{---T-A-G-C-G-A---}3' \\
3'\text{---A-T-C-G-C-T---}5'
\end{array}
$$

DENATURATION OF DNA

The double-helical form of DNA unwinds into a single-stranded disordered structure called random coils when subjected to: (1) extremes of pH; (2) increased temperature; (3) decrease in dielectric constant by alcohols, ketones, etc.; and (4) exposure to urea or amides. During this denaturation process, no covalent bonds in the backbone structure are broken.

Two types of forces maintain the double-helical structure of DNA: (1) hydrogen-bonding between base pairs and (2) hydrophobic interactions between successive stacked bases. When either or both sets of forces are interrupted, the double helix denatures into a random coil.

HYBRIDISATION OF DNA

In 1961 it was discovered that complementary single strands of DNA will re-form double helices in a process called DNA renaturation or hybridisation. In fact, under the appropriate conditions, hybridisation can occur between any two single-stranded nucleic acid chains (DNA : DNA, RNA : RNA or RNA : DNA) provided that they have a complementary nucleotide sequence.

Use of this technique of hybridisation is now very wide. By using cloned, single-stranded, radiolabelled DNA (commonly referred to as a DNA probe), one can determine the number and structure of any particular gene in the DNA of a cell and also the type and quantity of transcribed RNA. Early studies involved the use of reactions in solution, called solution hybridisation, but now the DNA or RNA under test is usually immobilised on a nitrocellulose or nylon filter, the so-called hybridisation of Southern (DNA) and Northern (RNA) blots.

The technique of solution hybridisation was based on the kinetics of nucleic acid hybridisation. Since the rate of hybridisation is limited by the rate at which two complementary nucleic acid chains happen to collide, the concentration of specific nucleotide sequences can be measured by the rate at which they hybridise to a radio-labelled cloned DNA of complementary sequence. This is such a stringent test that sequences present as a single copy in the genome can be detected. Such studies can also be used with RNA to determine whether cloned DNA sequences are transcribed into RNA, and if so how many copies of RNA are made in a cell. Use of more elaborate hybridisation procedures (the S_1 mapping technique) enables identification of the exact region of the cloned DNA that hybridises to RNA and thereby definition of start and stop sites for RNA transcription.

Nowadays, the DNA or RNA under test is usually immobilised on a filter and hybridisation to a single-stranded radiolabelled DNA probe is carried out on the filter itself so that the end result can be visualised by autoradiography. If the DNA or RNA fragments under test are initially separated by size on gel electrophoresis before transfer to a filter, the sizes of fragments containing nucleic acid sequences complementary to the cloned DNA probe can be determined. For RNA, this enables an instant estimation of the size of the transcribed RNA. If DNA is cut into a series of restriction fragments using different restriction enzymes before separation on the gel, a restriction map of the gene of interest can be built up. In this map, the relationship of each cutting site to its neighbours can be estimated (Fig. 1.4). Since such maps reflect the arrangement of selected nucleotide sequences, they are used for comparing normal with mutant or variant genes and also for studying homology between genes.

One of the major medical applications of restriction maps is in the diagnosis of genetic diseases. Sickle-cell anaemia results from a mutation that changes a glutamic acid residue to a valine residue at position 6 in the β-globin chain of haemoglobin. The mutation involves a single base change in the DNA. This disease was one of the first genetic diseases to be diagnosed directly at the gene level by restriction enzyme analysis of the DNA. The restriction enzyme MstII generates from normal DNA a 1.1 kilobase β-globin gene fragment, but in sickle-cell DNA this is replaced by a 1.3 kilobase fragment.

In many higher plant and animal DNAs, a significant fraction of the cytosine residues exist in a modified form in which a methyl group is attached to the 5 carbon atom of the pyrimidine ring (5-methylcytosine). It is now clear that these methyl groups are added after the DNA chains are synthesised and may play a key role in the control of DNA transcription. Analysis of restriction fragments produced by enzymes sensitive to DNA methylation (e.g. HpaII and MspI cut at the same site in DNA but differ in their sensitivity to cytosine methylation) can be useful in determining the role and position of DNA methylation in gene expression.

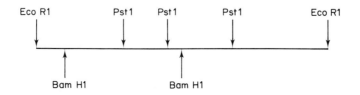

Fig. 1.4 A simple diagrammatic representation of a restriction map of a DNA fragment

DNA CLONING

In cloning procedures, a specific DNA fragment is integrated into a rapidly replicating genetic element (plasmid or bacteriophage) so that it can be amplified in bacteria or yeast cells.

There are two key features in the cloning of DNA: (1) the use of the cloning vector and (2) the combined use of restriction enzymes with DNA ligase.

1. Plasmids are small, circular, double-stranded DNA molecules that occur naturally in both bacteria and yeasts, where they replicate as independent units. Their features essential to DNA cloning are their small size (enabling simple purification) and the fact that they carry genes (which can be used for selection, such as resistance to antibiotics).

2. The combined use of restriction enzymes and DNA ligase has made it possible to graft and excise fragments of any DNA into these self-replicating elements. The staggered cleavage by restriction enzymes producing cohesive ends enables complementary base pairing between any two DNA fragments cut with the same enzyme. A circular DNA molecule, such as a plasmid, that is cut at a single site so it has cohesive ends, will tend to re-form a circle by annealing (base-pairing) of its cohesive ends. However, if a second piece of DNA cut with the same enzyme is mixed in, the two DNAs can join together. Once the ends are base-paired, they can be sealed with the enzyme DNA ligase (Fig. 1.5).

Thus, for DNA cloning, the plasmid DNA is cut with a restriction enzyme giving compatible cohesive or blunt ends to the DNA fragment to be cloned. The two DNA molecules are then joined (ligated) together using the enzyme DNA ligase. The hybrid plasmid is reintroduced into bacteria and those cells which have taken up the DNA are selected by the antibiotic resistance conferred on them by the plasmid. As these bacteria multiply, the plasmid also replicates to produce enormous numbers of copies of the original DNA fragment. The plasmid can be easily purified by virtue of its small size compared to the host cell DNA, and the copies of the original DNA fragment can then be recovered from the plasmid by excision with a second treatment with the same restriction enzyme.

The DNA to be cloned can be obtained either from cleavage of the entire genome of a cell with a specific restriction enzyme (genomic cloning) or from DNA synthesised from purified cellular mRNA using the enzyme reverse transcriptase (cDNA cloning). Cloned cDNA will contain mainly sequences coding for protein, but cloned genomic fragments may also contain DNA sequences 5′ or 3′ to the gene of interest or intervening sequences.

The ability to clone DNA and the ease with which DNA can now be analysed by hybridisation techniques and sequencing have had innumerable scientific and commercial applications. Cloned DNA can be used to search for DNA sequences or transcribed RNA in any cell, and comparisons can be made between normal and variant or abnormal situations. For example, in 1982 a cancer gene from human bladder cells was isolated and cloned in E. coli. The base sequence of this cancer gene was found to differ from its normal counterpart by only a single base change, leading to a single amino acid change in the protein product.

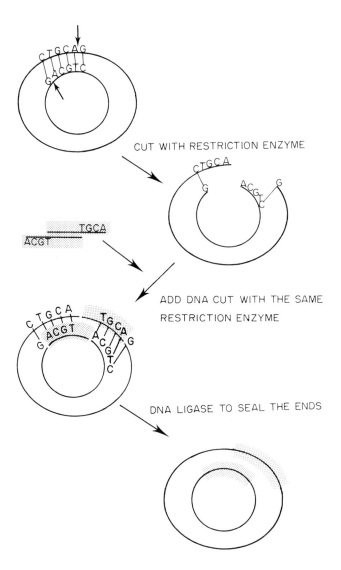

Fig. 1.5 Principles of cloning a piece of DNA into a plasmid molecule

Further advances have been made by reintroducing cloned DNA sequences back into cells. This can be done in animal cells by one of three methods: (1) microinjection; (2) using viral vectors; and (3) by the process of transfection, whereby DNA is taken up by cells as a calcium precipitate. In plant cells, this is done using the crown gall plasmids of certain soil bacteria; these plasmids naturally integrate into the chromosomes of the plant cells which the bacteria infect. Since it is now possible to create mutations *in vitro* in any cloned DNA or to join different pieces of cloned DNA together, the effects of these changes on cells or organisms *in vivo* can be analysed. By altering, for example, the sequences controlling gene transcription in cloned genomic DNA, much has been learnt about how genes are switched on and off. On the commercial side, DNA cloning offers the possibility of large-scale economical production of protein hormones and vaccines.

Chapter 2

Preparation of DNA from tissues or cells in culture

2.1 PRINCIPLES INVOLVED

Single cells in suspension are lysed with a detergent (sodium dodecylsulphate (SDS)) and incubated with the enzyme, proteinase K, to break down protein molecules. This aqueous solution is then extracted with phenol. Phenol and water are immiscible, and protein is extracted into the phenol layer leaving nucleic acids in the aqueous layer. By the addition of two volumes of ethanol to the aqueous phase, the high-molecular-weight nucleic acids are then precipitated as a white fibrous material. Precipitation as a colourless gelatinous material indicates that proteins are still bound to the nucleic acid. The nucleic acids are treated with deoxyribonuclease (DNase)-free ribonuclease (RNase) to remove RNA followed by a second round of proteinase K treatment, phenol extraction and ethanol precipitation. By this stage, the DNA should be a clean white fibrous material. Any proteins remaining attached to the DNA will prevent good digestion by restriction enzymes (see Chapter 3), and so, if gelatinous precipitates remain, they must be treated with more proteinase K and phenol extraction. These methods are described by Pellicer *et al.* (1978), and the principles are outlined in Fig. 2.1.

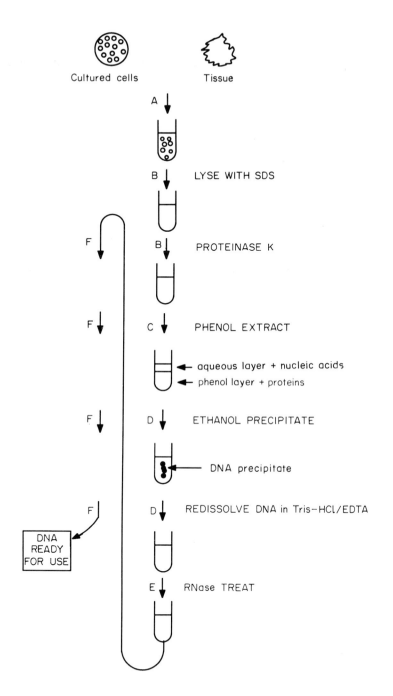

Fig. 2.1 Principles in the preparation of DNA

2.2   DNA FROM CULTURED CELLS: WEEKLY SCHEDULE

> MONDAY   Steps A–D (see Fig. 2.1)
> TUESDAY   Steps E–F (see Fig. 2.1)
> WEDNESDAY   DNA ready for use

2.3   DNA FROM CULTURED CELLS: LABORATORY METHODS

To avoid nuclease contamination, gloves should be worn throughout. All pipettes, glassware and solutions should be autoclaved before use. To avoid shearing high molecular weight DNA, any pipettes used should be of the wide-bore agar type (3 mm diameter). All quantities given are for preparation of DNA from 10^7 cells and should be scaled up as appropriate.

A.   Preparation of cell pellet

DNA can be made from cells growing in monolayer or in suspension culture, and in general, 100 μg DNA will be obtained from 10^7 cells. Cells growing in monolayer can be harvested either with enzyme treatment (e.g. trypsin or collagenase) or by scraping the cells gently off the dish with a rubber policeman. The latter treatment is easier and used for most purposes but the former treatment may be better for delicate primary cultures of cells. Cells growing in suspension in liquid or semisolid medium can be transferred to a centrifuge tube simply with a pipette. During harvesting, all cell preparations should be kept ice-cold.

> **Solutions and chemicals required:**
>   Phosphate-buffered saline (PBS)

> **Equipment recommended:**
>   Pipettes
>   Rubber policeman
>   Sterile plastic centrifuge tubes (Falcon)
>   Bucket of ice
>   Refrigerated centrifuge

PROTOCOL A

1. Cells growing in monolayer culture are washed twice on the dish with PBS. They are then scraped off the dish in cold PBS with a rubber policeman, transferred to a sterile plastic centrifuge tube on ice and pelleted by centrifugation at 4°C at 1000 g for 10 minutes.
2. Cells growing in suspension are transferred directly in cold PBS to a sterile plastic centrifuge tube on ice. The cells are pelleted by centrifugation at 4°C at 1000 g for 10 minutes and the cell pellet washed twice with ice-cold PBS.
3. The cell pellet can be used at once or stored frozen below −70°C.

B.   Lysis of cells

The cells are lysed with detergent (SDS) and incubated with proteinase K to break down protein molecules.

Solutions and chemicals required:
1 M Tris−HCl pH 7.4
0.5 M EDTA pH 7.5
1 mg/ml proteinase K (Boehringer) in water.
  This must be freshly made
Distilled water
10% (w/v) SDS (Serva) in water

Equipment recommended:
Pipettes
Water bath at 37°C

PROTOCOL B

1. The cell pellet is resuspended in an aliquot of the following solution, using 1.0 ml for 10^7 cells.

1 M Tris−HCl pH 7.4	10 μl
0.5 M EDTA pH 7.5	20 μl
1 mg/ml proteinase K	100 μl
Distilled water	870 μl

2. Add as fast as possible 50 μl of 10% SDS.
3. Incubate at 37°C in a water bath for 4 hours.

C. Phenol extraction

The lysate is extracted with phenol. Since phenol and water are immiscible, the protein extracted into the phenol layer is removed from nucleic acids left in the aqueous layer. Phenol used in this procedure must be very pure and should be redistilled before use, water-saturated and equilibrated with Tris buffer. Nowadays, redistilled phenol can be purchased from companies (such as Rathburn Chemicals) but any other phenol product must be redistilled before use. Some years ago, isoamyl alcohol was added to the phenol to prevent frothing but this is no longer used for this protocol since use of chloroform appears to be adequate. The final washes with chloroform alone serve to remove all traces of phenol.

Solutions and chemicals required:
5 M NaCl
Chloroform
Phenol (see below)

Preparation of phenol:
(i) Take 100 ml water-saturated redistilled phenol (Rathburn Chemicals).
(ii) Add 100 ml 0.5 M Tris−HCl pH 8.0 containing 1 mM EDTA.
(iii) Mix well and leave for 1 hour at room temperature.
(iv) Remove aqueous phase and discard.
(v) Repeat procedures (ii), (iii) and (iv) once more.
(vi) Add 100 ml 10 mM Tris−HCl pH 8.0 containing 1 mM EDTA.
(vii) Mix well and leave 1 hour at room temperature.
(viii) Remove aqueous phase and discard.
(ix) Repeat procedures (vi), (vii) and (viii) twice more.
(x) Add 100 ml 10 mM Tris−HCl pH 8.0.
(xi) Store at 4°C in the dark.

With time, phenol will oxidise and change from colourless to an orange colour. This can be slowed down by addition of 0.02% 8-hydroxyquinoline to the phenol. The solution can, however, be used for 6 months or more.

Phenol causes severe burns and must be handled with care.

Equipment recommended:
  Agar pipettes (wide-bore)
  Corex glass centrifuge tubes
  Bench-top centrifuge at room temperature

PROTOCOL C

1. Add 30 μl 5 M NaCl to each 1 ml of cell lysate.
2. Transfer to a Corex glass centrifuge tube.
3. Add an equal volume of phenol : chloroform (1 : 1 v/v) mixture.
4. Seal tube and mix gently by inverting the tube.
5. Separate the phases by low-speed centrifugation (2000 g for 10 min) at room temperature.
6. Remove the upper aqueous layer to a clean Corex glass centrifuge tube.
7. Repeat procedures 3–6 once more.
8. Add an equal volume of chloroform alone.
9. Seal tube and mix gently by inverting the tube.
10. Separate the phases by low-speed centrifugation (2000 g for 10 min) at room temperature.
11. Remove the upper aqueous layer to a clean Corex glass centrifuge tube.
12. Repeat procedures 8–11 once more.

D. Ethanol precipitation

High-molecular-weight nucleic acid can be precipitated from an aqueous salt solution by the addition of ethanol. Nucleic acid free of protein will be a white fibrous material. Precipitation as a colourless gelatinous material indicates the presence of protein bound to the DNA.

Solutions and chemicals required:
  100% ethanol
  70% v/v ethanol in water
  1 M Tris–HCl pH 8.0
  0.5 M EDTA pH 8.0
  Distilled water

Equipment recommended:
  Clear plastic disposable tubes
  Shaking water bath at 37°C

PROTOCOL D

1. To the final aqueous phase, add 2 volumes of 100% ethanol.
2. Swirl gently to mix and the nucleic acid precipitate should be visible at once.
3. Remove the precipitate by fishing it out on a pipette and wash by immersion in 70% ethanol and then 100% ethanol for a few seconds each. It is preferable not to subject the precipitate to centrifugation as this will compact it and make redissolving in water later very difficult.
4. Allow the precipitate to dry in the air for a few minutes, until it is no longer soaked in ethanol but not until it is bone dry.
5. Redissolve in the following solution, using 1.0 ml for each 10^7 cells at the beginning:

1 M Tris–HCl pH 8.0	10 μl
0.5 M EDTA pH 8.0	20 μl
Water	970 μl

   If the DNA does not dissolve immediately it should be incubated at 37°C with gentle rocking overnight. Sometimes it can take as long as 2 days to dissolve.

E. Ribonuclease treatment

Treatment so far will have resulted in isolation of all high-molecular-weight nucleic acids. RNA is removed by treatment with ribonuclease.

Solutions and chemicals required:
Stock ribonuclease A 51 mg/ml in water (stored at −20°C)
Stock ribonuclease T1 1000 units/ml in water (stored at −20°C)
10% (w/v) SDS (Serva) in water
Distilled water

Equipment recommended:
Water bath or oil bath at 95°C
Water bath at 37°C

PROTOCOL E

1. Take 2 μl of stock ribonuclease A and 5 μl of stock ribonuclease T1 and add 200 μl water. Heat to 95°C for 5 minutes. This destroys any DNase contaminant.
2. Add 40 μl of the treated ribonuclease to each 1 ml of DNA.
3. Incubate at 37°C in a water bath for 2 hours.
4. Add 40 μl 5 M NaCl and 10 μl of 10% SDS.

F.   Final steps

The final steps involve a second round of proteinase K treatment, phenol extraction and ethanol precipitation.

Solutions, chemicals and equipment required:
  As protocols B–D

PROTOCOL F

1. To mixture at the end of protocol E, add 100 μl proteinase K solution (1 mg/ml).
2. Incubate at 37°C in a water bath for 2 hours.
3. Repeat protocol C, steps 2–12 once.
4. Repeat protocol D, steps 1–4 once.
5. Redissolve the final DNA in the following solution, using 100 μl for each 10^7 cells at the beginning:

1 M Tris–HCl pH 8.0	1 μl
0.5 M EDTA pH 8.0	2 μl
Water	997 μl

   The DNA solution should be viscous with a concentration of about 1 μg/μl.
   If the DNA does not dissolve at once, it should be incubated at 37°C with gentle rocking for up to 2 days.

2.4 DNA FROM TISSUES: WEEKLY SCHEDULE

MONDAY	Steps A–B (see Fig. 2.1)
TUESDAY	Steps C–D (see Fig. 2.1)
WEDNESDAY	Steps E–F (see Fig. 2.1)
THURSDAY	DNA ready for use

2.5 DNA FROM TISSUES: LABORATORY METHODS

DNA is prepared from tissues in a way analogous to that described for DNA from cultured cells (Blin and Stafford, 1976). The main problem encountered here is that of disaggregating the cells in the tissue to produce a single-cell suspension. This can be done simply by grinding the frozen tissue with a pestle and mortar in liquid nitrogen. However, if it is available, a dismembranator makes life easier. The other variation is that it is often better to dialyse the DNA solution instead of ethanol precipitation at step D to avoid problems of redissolving. Ethanol precipitation, however, is still advisable at the end.

A. Disaggregation of cells in the tissue

Equipment recommended:
  Pestle and mortar or dismembranator
  Liquid nitrogen

PROTOCOL A

1. Freeze the tissue in dry ice or liquid nitrogen.
2. Grind the tissue to a fine powder. This can be done either by simply using a pestle and mortar, keeping the tissue frozen by covering in liquid nitrogen while grinding, or by using a dismembranator.

B. Lysis of cells

Solutions and chemicals required:
  Tissue buffer mix, made freshly:

1 M Tris–HCl pH 7.4	0.1 ml
0.5 M EDTA pH 7.5	0.2 ml
Proteinase K 1 mg/ml	1.0 ml
Distilled water	8.7 ml

  10% w/v SDS (Serva) in water

Equipment recommended:
Glass conical flask
Shaking water bath at 37°C

PROTOCOL B

1. Transfer tissue powder to a glass conical flask.
2. Add tissue buffer mix at room temperature: 10 volumes of buffer is adequate for most tissues but 40 volumes is advisable for spleen.
3. Add 0.5 volume of 10% SDS.
4. Incubate at 37°C in a shaking water bath overnight.

C. Phenol extraction

As for cultured cells (see section 2.3, protocol C).

D. Dialysis of DNA solution

Rather than attempting to precipitate the DNA in ethanol and risking problems in redissolving the DNA, it may be better to dialyse the DNA solution at this stage.

Solutions and chemicals required:
Dialysis solution made as follows:

1 M Tris−HCl pH 8.0	200 ml
0.5 M EDTA pH 8.0	80 ml
5 M NaCl	8 ml
Distilled water	Make volume to 4 litres

Equipment recommended:
5-litre beaker
Dialysis tubing

PROTOCOL D

1. Transfer DNA solution to dialysis tubing.
2. Dialyse against 4 litres of dialysis solution over 24 hours, changing the solution at least 4 times.

E.   Ribonuclease treatment

As for cultured cells (see section 2.3, protocol E).

F.   Final steps

As for cultured cells (see section 2.3, protocol F).

2.6   CONDITION OF DNA

1. The concentration of the DNA can be measured using optical density.
   Take 2 μl of the DNA solution and add 500 μl of water.
   Read optical density at 260 nm (OD_{260}).
   A solution of DNA in water of 1 mg/ml gives an OD_{260} of 20.
   Thus $\mu g/\mu l$ in DNA solution $= 1/20 \times OD_{260} \times 250$.
2. When it is prepared, DNA must be of high molecular weight and not broken down into small fragments. This must be ensured for each DNA sample and can be checked simply by running a small aliquot on an 0.8% agarose gel. This is done by following protocol C in section 3.3 of Chapter 3. 1 μg of DNA is loaded into a well of a gel and run as a fast gel. Molecular weight markers provided by HindIII cut phage λ DNA should be run alongside. If it is high molecular weight, the DNA should run no further than the slowest of the marker bands (which is 23 000 nucleotides in size).

2.7   REFERENCES

Blin, N. and Stafford, D.W. (1976) *Nucleic Acids Res.* **3:** 2303–2308.
Pellicer, A., Wiger, M., Axel, R. and Silvestein, S. (1978) *Cell* **14:** 133–141.

Chapter 3

DNA analysis by restriction-enzyme digestion and Southern blotting

3.1 PRINCIPLES INVOLVED

DNA analysis using restriction enzymes and Southern blotting enables the presence and structure of genes in cells to be studied in some detail fairly simply and quickly. The principles involved are illustrated in Fig. 3.1. High-molecular-weight DNA is cleaved with one or more restriction endonucleases, which cut DNA in a very site-specific manner. The resulting DNA fragments are then separated by size on agarose gel electrophoresis. The double-stranded DNA fragments are denatured in the gel and the single-stranded pieces can then be transferred to a sheet of nitrocellulose. It is this transfer of electrophoretically-resolved DNA fragments to nitrocellulose filters which is known as 'Southern blotting' (Southern, 1975). This entails laying a sheet of nitro-cellulose (which acts as a filter) on top of the gel and establishing a flow of buffer through the gel and the nitrocellulose filter. The buffer carries the DNA fragments upwards from the gel to the nitrocellulose, where they subsequently bind. A single-stranded nucleic acid probe specific for the gene under study is then radiolabelled and hybridised to the filter. This probe can be a purified RNA, cDNA or cloned fragment of genomic DNA. Whichever is chosen, the labelled probe will hybridise to any DNA fragment on the filter which contains complementary nucleotide sequences. Autoradiography of the nitrocellulose filter will reveal the position of each piece of DNA that contains any part of the probe used. By comparing the positions with radiolabelled polynucleotide markers of known length, the molecular size of each band can be estimated.

Several parameters can be analysed from these restriction-enzyme patterns:

(1) The number of copies of a gene can be estimated by comparing relative intensities of bands on the X-ray film.
(2) By using different combinations of restriction enzymes and different parts of the probe, a precise 'restriction-enzyme map' can be established.
(3) Variant genes sometimes differ from their normal counterparts by the loss or gain of certain restriction-enzyme sites or in the size of fragments generated. Such alterations can occur within either the structural gene or surrounding control areas, and can either be related directly to the abnormality or result from a polymorphism with linkage to the abnormal gene. Such variations have already proved useful in the diagnosis of certain clinical syndromes.

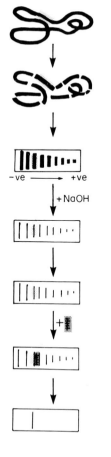

High-molecular-weight DNA

A,B Cut DNA with restriction
     enzyme(s)

C   Separate DNA fragments
     on agarose gel

−ve ⟶ +ve

+NaOH

D   Denature DNA to single strands

D   Transfer DNA to nitrocellulose
     filter

E,F Hybridise single-stranded DNA
     on filter to single-stranded
     labelled DNA or RNA probe ▮

G   Wash filter to remove non-
     specific hybridisation and
     expose to X-ray film

Specific DNA fragments hybridised
to the probe give a band on
autoradiography

Fig. 3.1 Principles of Southern blotting

3.2 WEEKLY SCHEDULE

MONDAY	Restriction-enzyme digestion overnight (Protocol A)
TUESDAY	Precipitate samples with ethanol (Protocol B)
	Run test gel (to test digestion is complete) (Protocol C)
	(Redigest samples if needed) (Return to Protocol A)
	Make main gel and run overnight (Protocol C)
WEDNESDAY	Denature DS DNA into SS DNA (Protocol D, 1–5)
	Blot DNA onto nitrocellulose overnight (Protocol D, 6)
THURSDAY	Disassemble blot and bake (Protocol D, 7)
	Nick translate DNA to make ^{32}P-probe (Protocol E)
	Hybridise blot to probe overnight (Protocol F)
FRIDAY	Wash blot, dry blot and put onto autoradiography (Protocol G)

3.3 LABORATORY METHODS

Nuclease contamination must be avoided in procedures A, B and E below. For these protocols, gloves should be worn throughout. Pipette tips, glassware and solutions should all be autoclaved before use. These precautions are not necessary for the other protocols.

A. Digestion of DNA with restriction endonucleases

The first step involves cleavage of high molecular weight DNA with one or more restriction endonucleases. These enzymes are now available from many commercial sources.

Equipment recommended:
Microcentrifuge tubes (1.5 ml volume (autoclaved) (Treff Lab)
Water bath

Solutions and chemicals required:
DNA prepared in water (see Chapter 2).
10 × restriction enzyme (10 × RE) buffer,
made as follows:
  2 ml 1 M Tris−HCl pH 7.4 (autoclaved)
  1.6 ml distilled water (autoclaved)
  0.4 ml 1 M magnesium chloride (autoclaved)
  28 μl stock β-mercaptoethanol
(This can be stored at −20°C for up to one
month.)
Restriction-enzyme preparation (Boehringer
or Biolabs)
5 M sodium chloride (autoclaved)
Distilled water (autoclaved)

PROTOCOL A

Set up the restriction-enzyme reaction mixture in a 1.5 ml
microcentrifuge tube:

1. Add up to 20 μg of DNA, prepared as in Chapter 2.
   DNA solutions are viscous and should be measured
   with a pipette tip cut to give wider bore (about 2 mm).
2. Add 5 μl of 10 × RE buffer.
3. Add the required amount of enzyme (up to 5 μl
   volume only). The amount of enzyme required is
   calculated in units (1 unit of enzyme cuts 1 μg of DNA
   in 1 hour at 37°C in a volume of 50 μl).
4. Add 5 M NaCl as needed by the enzyme.
   There are three groups of enzyme:
     Low-salt enzymes   − add no 5 M NaCl
     Medium-salt enzymes − add 0.5 μl 5 M NaCl
     High-salt enzymes   − add 1.0 μl of 5 M NaCl
   The salt requirements are specified by the supplier.
5. Add distilled water to make the volume up to 50 μl.
   The ideal volume for the reaction is 50 μl. However,
   larger volumes can be used if more than 20 μg DNA
   is to be cut or if the enzyme is so dilute that more than
   5 μl has to be used. It should be remembered that the
   volume of enzyme used should never exceed one-
   tenth of the reaction volume.
6. Incubate reaction mixture for 1 hour minimum. Most
   enzymes work best at 37°C but some require other
   temperatures. This will be specified by the supplier.
   Theoretically, 1 hour should be adequate, but it is often
   both convenient and advantageous to leave the
   reaction overnight.

B.   Ethanol precipitation

After restriction-endonuclease digestion, the samples are precipitated with ethanol. This serves to remove salts which might interfere during subsequent gel electrophoresis and also to concentrate the DNA into a smaller volume for loading onto the gel.

Solutions and chemicals required:
   5 M sodium chloride (autoclaved)
   100%, 95%, 70% ethanol (v/v in water)
   Distilled water (autoclaved)

Equipment recommended:
   Flask of dry ice
   Microcentrifuge (Eppendorf or M.S.E.)
   Spectrophotometer

PROTOCOL B

1. Make 0.1 M with NaCl [i.e. add 1 μl of stock 5 M NaCl (autoclaved) to each 50 μl of digest].
2. Add 2.5 vol. of 95% ethanol.
3. Leave in dry ice for 30 minutes.
4. Spin in microcentrifuge at 4°C for 10 minutes.
5. Pour off ethanol and drain well from DNA pellet.
6. Add 0.5 ml of 70% ethanol (dissolves and removes NaCl).
7. Spin in microcentrifuge at 4°C for 10 minutes.
8. Again pour off ethanol and then add 0.5 ml of 100% ethanol (to dry).
9. Spin in microcentrifuge at 4°C for 10 minutes.
10. Pour off ethanol. Dry DNA pellet in air.
11. Take up DNA in distilled water (so that DNA is about 1 μg/μl).
12. Measure concentration of DNA:
      Take 2 μl of the DNA and add to 500 μl of water.
      Read optical density at 260 nm (OD_{260}).
      A solution of DNA in water 1 mg/ml gives an OD_{260} of 20.
      Thus μg/μl in DNA solution = 1/20 × OD_{260} × 250.

C.   Prepare and run an agarose gel for DNA

The DNA fragments produced from restriction-enzyme digestion are separated by size on an agarose gel.

Solutions and chemicals required:
   5 × TBE buffer, made as follows:
      106 g Tris base
      55 g Boric acid
      9.3 g dipotassium EDTA
      Dissolve and make up to 2 litres with water
   10 mg/ml ethidium bromide (BDH) in water
      (store in the dark)
   Agarose powder (SEAKEM-ME) (FMC Corp.)
   DNA sample buffer (DNA SB) made as
   follows:
      330 μl glycerol
      75 μl 5 × TBE buffer
      250 μl 0.5 M EDTA (pH 7.0)
      Mix thoroughly, then
      Add 27 μl of 10% w/v SDS (Serva) in water
      (if this is added earlier it precipitates out)
      Add bromophenol-blue powder (spatula
      tip, approx. 100 μg)

Equipment recommended:
   Microwave oven or boiling water bath
   Agarose gel electrophoresis equipment
      (commercially available or home-made)
   Electrophoresis power supply

PROTOCOL C

1. Prepare mould for the gel.
   Many systems are available commercially for pouring
   gels but a simple home-made system is illustrated in
   Fig. 3.2.

A perspex plate is levelled using a spirit level.
A glass plate (14 cm × 19 cm) is used as the mould.
Its edges are sealed with PVC electrical tape (or
radioactive/biohazard-labelled tape) so that it forms a
wall 1 cm high all around the plate. This is placed on
the level perspex plate. A simple perspex comb is
assembled at one end of the plate so that there is about
1 mm of space between the base of the teeth and the
glass plate.

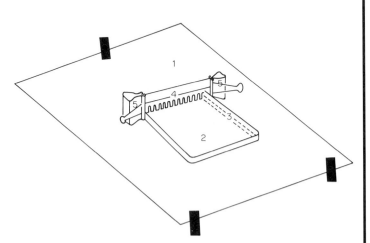

Fig. 3.2 Mould for pouring agarose gels.
1 = perspex plate levelled with three screw feet;
2 = glass plate 14 cm × 19 cm; 3 = PVC tape;
4 = perspex comb to mould wells in the gel;
5 = clamp to hold comb in position

2. Weigh 0.8 g of agarose into a 200 ml conical flask.
   Add 100 ml of 1 × TBE buffer.
   Add 100 μl ethidium bromide solution (10 mg/ml in
   water).
   Heat in microwave oven or boiling water bath until
   agarose dissolves.
3. Pour agarose onto gel mould and allow to set (1–2
   hours).
   NB The recipe given here provides a 0.8% agarose
   gel, which is useful for most purposes. To separate
   very large or very small DNA fragments, the concentration can be varied. (For further details see Maniatis
   et al., 1982.)
4. The gel tank is prepared and connected to an electrophoresis power supply. Many systems are available
   commercially but a simple home-made system is
   illustrated in Fig. 3.3.

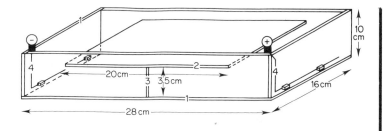

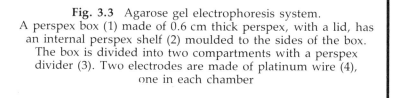

Fig. 3.3 Agarose gel electrophoresis system.
A perspex box (1) made of 0.6 cm thick perspex, with a lid, has
an internal perspex shelf (2) moulded to the sides of the box.
The box is divided into two compartments with a perspex
divider (3). Two electrodes are made of platinum wire (4),
one in each chamber

5. Place 1600 ml of 1 × TBE buffer into the gel tank and
   add 2 drops of ethidium bromide solution (10 mg/ml).
6. Put the gel into the tank. The surface of the gel should
   be just submerged (by about 1 cm depth of buffer).
7. Make DNA samples up to 20 μl in water and add 8 μl
   of DNA SB.
8. Load DNA into wells of the gel.
   Polynucleotides of known size can be loaded as
   required in a separate well, to provide molecular
   weight markers (see section 3.4).
9. A fast test gel should be run initially to assess if the
   DNA is well cut:
   Load 1 μg of DNA per well.
   Run 120 V (50 mA) 2–4 hours.
   If DNA is not cut to completion, it must be redigested
   with more enzyme (return to protocol A). If DNA is
   cut, the fragments can be separated on a main gel for
   Southern blotting. This would be run as follows:
   Load up to 20 μg DNA per well.
   Run 25–30 V (20 mA) 18 hours.
   All samples are run from the negative to the positive
   electrode.

D.   Southern transfers (blotting)

This procedure was first described by Southern (1975). It involves transfer of the electrophoretically resolved DNA fragments from the agarose gel onto a nitrocellulose filter.

Solutions and chemicals required:
  1 M sodium hydroxide
  5 M sodium chloride
  1 M Tris–HCl pH 7.4
  Distilled water
  10 × SSC solution made as follows:
    87 g sodium chloride
    44.1 g trisodium citrate dihydrate
    Dissolve in 800 ml distilled water
    Adjust pH to 7.0 with 5 M NaOH
    Make volume up to 1 litre with water
  Nitrocellulose. This is available from many commercial sources (recommended: Sartorius filters type SM26; pore size 0.1 μm).

Equipment recommended:
  Ultraviolet light
  2 plastic trays (25 cm × 30 cm × 5 cm high)
  2 large rubber bungs (size 49)
  2 glass plates (14 cm × 19 cm)
  Whatman filter paper (No. 1 and No. 3mm)
  One packet of paper hand towels
  1 kilogram weight (e.g. a 500-ml bottle filled with water)
  Clingfilm
  Vacuum oven

PROTOCOL D

1. DNA in the gel is stained with the ethidium bromide in the buffer and can be visualised under ultraviolet light. The DNA bands in the agarose gel are thus examined firstly under ultraviolet light, and photographed for a permanent record if a camera is available.
2. Place the gel in a plastic tray of convenient dimensions.
3. Denature the DNA into separate strands by using 2 × 30 minute washes of 250 ml each of the solution made as indicated below:
   250 ml 1 M NaOH
   100 ml 5 M NaCl
   150 ml distilled water
4. Neutralise the gel with 2 × 30 minute washes of 250 ml each of the solution made as indicated below:
   250 ml 1 M Tris−HCl pH 7.4
   250 ml 5 M NaCl
5. Soak the gel in 10 × SSC solution for a few minutes.
6. The blot is set up as shown in Fig. 3.4.
   (a) Wet the nitrocellulose filter in distilled water in a plastic tray, then transfer to 10 × SSC solution.
   (b) In another plastic tray, place two large rubber bungs on the bottom.
   (c) Place a glass plate (14 cm × 19 cm) on top.
   (d) Now place a sheet of 3mm Whatman paper over the glass plate so that it acts as wicks on either side of the plate.
   (e) Add 600 ml of 10 × SSC solution to the tray and soak the Whatman paper, making sure that all air bubbles are removed.
   (f) Carefully place the gel on top of the Whatman paper, again removing all air bubbles.
   (g) Now cover all edges of the gel using four pieces of clingfilm, tucking excess under the tray but taking care not to cover any areas of the gel containing DNA. This ensures that the 10 × SSC solution passes evenly through the gel.
   (h) Lay the wetted nitrocellulose sheet onto the gel surface, taking care to exclude air bubbles.
   (i) Lay three sheets of presoaked (10 × SSC) Whatman No. 1 paper over the filter.
   (j) Place about 8 cm of dry Kleenex Hi-Dri paper towels over the Whatman No. 1 paper and place a glass plate on top of this. Now put a kilogram weight on top and leave to blot overnight (a minimum of 18 hours).

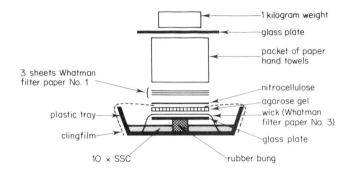

Fig. 3.4 Diagram of apparatus for the Southern blot

7. Disassemble blot the next morning.
   Place the filter (TAKE CARE: with DNA upwards!) on
   dry tissues.
   Leave to dry in air for about 30 minutes.
   Check the gel under ultraviolet light to ensure that all
   the DNA has transferred.
   Place the dried filter in a small envelope of filter paper
   (Whatman No. 3mm) and bake in a vacuum oven at
   80°C for 3–5 hours.

E. To make ^{32}P-labelled probe by nick translation of DNA

Radiolabelled probes for nucleic acid hybridisation can be prepared by a variety of
methods, which are discussed in more detail elsewhere (Arrand, 1985). In this protocol,
one reliable method only is presented. This involves labelling the probe by a nick translation reaction (Rigby *et al.*, 1977). The enzyme DNase I is used to create single-strand
nicks in double-stranded DNA. Then the $5' \rightarrow 3'$ exonuclease and $5' \rightarrow 3'$ polymerase
actions of *E. coli* DNA polymerase I are used to incorporate radiolabelled deoxyribonucleotides while repairing the nicks. Finally, the DNA is denatured with sodium
hydroxide and free unincorporated deoxyribonucleotides are separated from the DNA
by column chromatography. In practice, the end result of hybridisation is usually cleaner
if DNA used in this reaction is a piece of DNA under 2000 bases in length and separated
from any plasmid vector used for cloning (see Chapter 6).

Solutions and chemicals required:
Distilled water (autoclaved)
$10 \times$ nick translation ($10 \times$ NT) buffer, made
as follows:
  500 μl 1 M Tris—HCl pH 8.0 (autoclaved)
  100 μl 0.5 M magnesium chloride (autoclaved)
  10 μl β-mercaptoethanol
  390 μl distilled water (autoclaved)
20 μM dATP, dGTP and dTTP, made as follows:
  Each deoxyribonucleotide triphosphate is
  dissolved in sterile distilled water to give
  a concentration of 10 mM. The pH of each
  solution is then adjusted to pH 7.0 using a
  solution of 0.05 M Tris base. The concentration of each solution is checked by reading
  the optical density of a small aliquot diluted
  in water (the molar extinction coefficient for
  the bases is:
    A $1.54 \times 10^4 \, M^{-1}cm^{-1}$ at 259 nm;
    G $1.37 \times 10^4 \, M^{-1}cm^{-1}$ at 253 nm;
    T $7.4 \times 10^3 \, M^{-1}cm^{-1}$ at 260 nm).
  These 10 mM stock solutions can be stored
  at $-20°C$ and diluted in water for use at
  20 μM concentration.
^{32}P-dCTP purchased from Amersham (PB10205)
DNA probe (prepared as described in
Chapter 6)
1 mg/ml bovine serum albumin (BSA) (Sigma,
Fraction V powder) in autoclaved distilled water
DNase I (Worthington). A stock solution of
  100 μg/ml can be stored at $-20°C$. This
  should be diluted to 100 ng/ml for use but
  at this lower concentration continual thawing and refreezing is not recommended.
  Aliquots should be stored frozen and
  thawed once only before discarding.
DNA polymerase I (Boehringer)
5 M sodium hydroxide (autoclaved)
G50 column buffer, made as follows:
  10 ml 5 M sodium chloride (autoclaved)
  5 ml 1 M Tris—HCl pH 7.4 (autoclaved)
  5 ml 10% w/v SDS (Serva) in water (filter-
    sterilised)
  480 ml distilled water (autoclaved)
Sephadex G50 fine (Pharmacia) pre-swollen
in the G50 column buffer

Equipment recommended:

^{32}P-dCTP emits beta-radiation and should be handled with care:

all work should be done behind perspex screens (1 cm thick) to protect the operator from radiation. Usual precautions for handling ^{32}P-labelled chemicals must be observed at all times.

Geiger counter (for beta-emission)

Microcentrifuge tubes 1.5 ml volume
  (autoclaved) (Treff Lab)

Water bath at 14°C (use a water bath in a cold room)

5 ml sterile disposable plastic pipette

Siliconised sterile glass wool

Sterile tubing to set up a chromatography column

Scintillation counting facilities

PROTOCOL E

Prepare nick translation mix in a 1.5 ml microcentrifuge tube (this should be done behind perspex screens):

1. 5 μl dATP (20 μM).
2. 5 μl dTTP (20 μM).
3. 5 μl dGTP (20 μM).
4. 5 μl ^{32}P-dCTP (50 μCi/reaction).
5. 0.5 μg DNA (volume 0.5 μl).
6. 24.5 μl water.
7. 5 μl 10 × NT buffer.
8. Add 1 μl of 1 mg/ml BSA to stabilise the DNase which is added later.
9. Add 2.5 μl of DNase I (100 ng/ml). Leave at room temperature for 3 minutes.
10. Add 1 μl of DNA polymerase I.
11. Leave at 14°C for 2 hours.
12. Add 3.3 μl of 5 M NaOH to the mix, and leave at room temperature for 10 minutes.

13. Separate ^{32}P-labelled DNA from ^{32}P-dCTP on a column of Sephadex G50 fine.

   (a) Prepare a column of Sephadex G50 fine in a disposable 5 ml pipette plugged with siliconised sterile glass wool. All column equipment should be sterile. Equilibrate the column with column buffer.

   (b) Apply the reaction mix to the column and elute with column buffer. Monitor effluent with a Geiger counter.

   (c) Two peaks are formed on the column. The first peak is the labelled DNA. The second peak is ^{32}P-dCTP. The first peak is collected in a volume of 1−2 ml.

   (d) Count a 5 μl aliquot by liquid scintillation counting to estimate the specific activity (cpm/μg) of the DNA.

F. Hybridisation

Single-stranded DNA immobilised on the nitrocellulose filter is hybridised to the single-stranded radiolabelled DNA probe. Several methods are available for this reaction but, commonly, formamide or dextran sulphate are used. Dextran sulphate, although viscous to handle, provides a reliable and sensitive assay method and is the method of choice described here. The procedure described here provides high stringency and only closely homologous DNA sequences will hybridise. If the DNA probe is unlikely to be closely related to the DNA under examination, the stringency can be lowered by either lowering the temperature or increasing the salt concentration (i.e. amount of 20 × SSC solution added).

Equipment recommended:

Plastic bags and a plastic-bag sealer (any commercially available system sold for domestic use is suitable)

An incubate set at 65°C with shaking facilities. (If this is not available, bags can be submerged in a shaking water bath.)

Solutions and chemicals required:
  Distilled water
  20 × SSC solution made as follows:
    174 g sodium chloride
    88.2 g trisodium citrate dihydrate
    Dissolve in 800 ml distilled water
    Adjust pH to 7.0 with 5 M NaOH
    Make volume up to 1 litre with water
  25 × Denhardt's solution, made as follows:
    2 g bovine serum albumin (BSA) (Sigma,
      Fraction V)
    2 g Ficoll 400 (Pharmacia)
    2 g polyvinyl pyrrolidone (Sigma)
    Make up to 400 ml with water
    Filter-sterilise by passing through a 0.45 μm
      filter
  10% w/v SDS (Serva) in water. Sodium
    dodecylsulphate (Serva) is dissolved in
    water (10 g/100 ml) and filter-sterilised
    (0.2 μm)
  Single-stranded (SS) DNA solution, made as
  follows:
    1 mg/ml calf thymus DNA (Sigma); dissolve
    in boiling water,
    sonicate (amplitude low, 1 minute), then
    boil again;
    store in a sterile bottle at 4°C
  50% dextran sulphate solution, made as
  follows:
    500 g (Pharmacia sodium salt), plus 800 ml
    water
    Shake occasionally by hand over several
    days to dissolve
    Make volume to 1 litre with water

PROTOCOL F

1. Prepare solution A:
   - 3 ml 20 × SSC solution
   - 8 ml 25 × Denhardt's solution
   - 9 ml distilled water

   ─────

   20 ml

   ─────

   Place the nitrocellulose blot inside a plastic bag. Add solution A. Expel air bubbles. Seal bag. Incubate for 1 hour at 65°C with vigorous shaking.
2. Prepare solution B:
   - 1 ml 20 × SSC solution
   - 8 ml 25 × Denhardt's solution
   - 0.2 ml 10% SDS solution
   - 1 ml SS DNA 1 mg/ml
   - 10 ml distilled water

   ─────

   20 ml

   ─────

   Cut off the corner of the plastic bag. Pour out solution A. Add solution B. Reseal. Incubate for 1 hour at 65°C with vigorous shaking.
3. Prepare solution C:
   - 1 ml 20 × SSC solution
   - 8 ml 25 × Denhardt's solution
   - 0.2 ml 10% SDS solution
   - 1 ml SS DNA 1 mg/ml
   - 6 ml distilled water
   - 4 ml 50% dextran sulphate

   ─────

   20 ml

   ─────

   Cut off the corner of the plastic bag. Pour out solution B. Add solution C. Reseal. Incubate for 1 hour at 65°C with vigorous shaking.
4. Prepare solution D:
   This is identical to solution C but contains in addition 5−10 million cpm of ^{32}P-labelled DNA probe.
   Cut off the corner of the plastic bag. Pour out solution C. Add solution D. Reseal. Incubate for 15−18 hours at 65°C with vigorous shaking.

G. Washing the blot

After hybridisation, the Southern blot is washed to remove nonspecific hybridisation. The important step in this procedure is step 3, which determines the stringency required. The conditions described here provide a high stringency wash. If the DNA probe is not closely related to the DNA under examination, the stringency of the wash in step 3 should be lowered by either decreasing the temperature or increasing the salt concentration (i.e. amount of 20 × SSC added).

Solutions and chemicals required:
  20 × SSC solution made as follows:
    174 g sodium chloride
    88.2 g trisodium citrate dihydrate
    Dissolve in 800 ml distilled water
    Adjust pH to 7.0 with 5 M NaOH
    Make volume up to 1 litre with water
  25 × Denhardt's solution, made as follows:
    2 g bovine serum albumin (Sigma)
    2 g Ficoll 400 (Pharmacia)
    2 g polyvinylpyrrolidone (Sigma)
    Make up to 400 ml with water
    Filter-sterilise at 0.45 μm
  10% w/v SDS (Serva) in water and filter-
    sterilised at 0.2 μm
  Distilled water

Equipment recommended:
  An incubator set at 65°C with shaking facili-
  ties (if this is not available, washing can be
  done in a shaking water bath)
  Plastic tray
  Clingfilm
  Whatman filter paper (No. 3mm)
  X-ray film cassette
  X-ray film (recommended: Kodak XAR5)
  Facilities for developing X-ray film

PROTOCOL G

1. Remove the nitrocellulose filter from the plastic bag in which hybridisation has taken place and put the filter into a plastic tray.
2. Wash at 65°C with 2 × 10 minute followed by 2 × 30 minute washes, 100 ml each time of solution made as indicated below:

      20 ml 20 × SSC solution
     160 ml 25 × Denhardt's solution
       4 ml 10% SDS solution
     216 ml distilled water
   _____

     400 ml
   _____

3. Wash at 65°C with 2 × 30 minute washes of 100 ml each of solution made as indicated below:

      2 ml 20 × SSC solution
      2 ml 10% SDS solution
    196 ml distilled water
   _____

    200 ml
   _____

4. Wash the blot with 4–5 rinses at room temperature (1 minute each) with 100 ml each of solution made as indicated below:

     75 ml 20 × SSC solution
    425 ml distilled water
   _____

    500 ml
   _____

5. While the blot is still wet it should be scanned with a Geiger counter. If a large amount of radioactivity is detected scattered all over the filter, there may be nonspecifically hybridised DNA still bound and the blot may need to be washed for longer (repeat steps 3 and 4).
6. After washing, allow the blot to air dry. Then place on 3mm Whatman paper and cover in clingfilm to prevent the X-ray film sticking to the blot. Expose to X-ray film.

3.4 RADIOLABELLING OF DNA MOLECULAR WEIGHT MARKERS

In order to determine the size of unknown restriction fragments, it is necessary to run, on the same gel, radiolabelled polynucleotide markers of known molecular weight. The markers used most frequently are derived from phage λ DNA cut with the enzyme HindIII, plasmid pAT 153 DNA cut with HinfI or plasmid pAT 153 DNA cut with HpaII. The sizes of the fragments generated from these digests are given in Table 3.1.

Table 3.1 Sizes of some useful molecular weight markers

DNA	Cut with enzyme	Fragments generated (in kilobases)
phage λ	HindIII	23,9.4,6.6,4.3,2.3,2.0,0.56,0.1
pAT 153	HinfI	1.63,0.517,0.396,0.298,0.221, 0.220,0.154,0.145,0.075
pAT 153	HpaII	0.622,0.492,0.404,0.242,0.238, 0.217,0.201,0.190,0.160,0.147, 0.122,0.110,0.090,0.076,0.067, 0.034,0.026,0.015,0.009

The polynucleotide markers are radiolabelled with ^{32}P-labelled phosphate at the 5' end. This is done by removal of the 5' phosphate group with the enzyme, calf intestinal phosphatase, and then adding back a 5' ^{32}P-labelled phosphate group with the enzyme T4 polynucleotide kinase.

Laboratory methods

Nuclease contamination must be avoided in this procedure. Gloves should be worn. Pipette tips, glassware and solutions should be autoclaved before use.

Equipment recommended:
Microcentrifuge tubes 1.5 ml volume (auto-
    claved) (Treff Lab)
Pipette tips (autoclaved)
Water baths at 37°C and 75°C
Microcentrifuge (M.S.E. or Eppendorf)
Dry ice
Scintillation counting facility for ^{32}P

Solutions and chemicals required:
  Phage λ or pAT 153 DNA
  Appropriate restriction enzyme
  10 × restriction-enzyme buffer (as Section 3.3, protocol A)
  5 M NaCl (autoclaved)
  Distilled water (autoclaved)
  Phenol (as Section 2.3, protocol C)
  Chloroform
  70% v/v ethanol in water
  100% ethanol
  3 M sodium acetate pH 5.2
  0.5 M EDTA pH 7.0
  10 × phosphate buffer made as follows:

1 M Tris pH 9.5	100 μl
100 mM spermidine	100 μl
0.5 M EDTA	2 μl
Distilled water	800 μl

  Calf intestinal phosphatase (Boehringer)
  10 × kinase buffer made as follows:

1 M Tris pH 9.5	500 μl
1 M magnesium chloride	100 μl
0.5 M dithiothreitol	100 μl
100 mM spermidine	50 μl
0.5 M EDTA	1 μl
Distilled water	249 μl

  γ-^{32}P-ATP (Amersham) (10 mCi/ml)
  T4 polynucleotide kinase (Boehringer)

PROTOCOL

1. Cut 5 μg of the appropriate DNA with the required enzyme in a reaction volume of 50 μl, following protocol A, Section 3.3.
2. Add 25 μl of phenol and 25 μl of chloroform. Vortex to mix and separate the two phases by centrifugation in a microcentrifuge for 5 minutes.
3. Place the upper aqueous phase in a clean microcentrifuge tube. Add 50 μl of chloroform and vortex to mix. Separate the two phases by centrifugation in a microcentrifuge for 5 minutes.
4. Place the upper aqueous phase in a clean microcentrifuge tube. Add 5 μl of 3 M sodium acetate pH 5.2 and 110 μl of 100% ethanol. Mix by gentle vortexing and freeze in dry ice for 30 minutes.
5. Spin down the DNA precipitate in a microcentrifuge on high speed for 15 minutes.
6. Pour off the ethanol. Add 50 μl of 70% ethanol and vortex. Spin down the DNA precipitate in a microcentrifuge for 10 minutes.
7. Pour off the ethanol. Add 50 μl of 100% ethanol and vortex. Spin down the DNA precipitate in a microcentrifuge for 10 minutes.
8. Pour off the ethanol and dry the DNA pellet in air.
9. Redissolve the DNA in 20 μl of distilled water. Add 3 μl of 10 $\times$ phosphatase buffer. Add the required number of units of calf intestinal phosphatase as specified by the manufacturer and make the volume up to 30 μl with distilled water.
10. Incubate in a water bath at 37°C for 60 minutes.
11. Incubate in a water bath at 75°C for 60 minutes to activate the phosphatase enzyme.
12. Add 5 μl of 10 $\times$ kinase buffer.
13. Move tube behind perspex screens for protection from radiation during all further steps of this protocol. Standard precautions must be taken for working with ^{32}P and all solutions and plastic ware must be disposed of in radiation bins. Add 100 μCi γ-^{32}P-ATP (10 μl of γ-^{32}P-ATP at 10 mCi/ml).
14. Add the required number of units of T4 polynucleotide kinase as specified by the supplier.
15. Incubate in a water bath at 37°C for 60 minutes, making sure the water bath is shielded from radiation.
16. Add 50 μl of phenol. Vortex to mix, and separate the two phases by centrifugation in a microcentrifuge for 5 minutes.

17. Place the upper aqueous phase in a clean micro-centrifuge tube. Add 25 μl of phenol and 25 μl of chloroform. Vortex to mix, and separate the two phases by centrifugation in a microcentrifuge for 5 minutes.

18. Place the upper aqueous phase in a clean micro-centrifuge tube. Add 50 μl of chloroform. Vortex to mix, and separate the two phases by centrifugation in a microcentrifuge for 5 minutes.

19. Place the upper aqueous phase in a clean micro-centrifuge tube. Add 5 μl of 3 M sodium acetate pH 5.2 and 2.5 μl of 0.5 M EDTA pH 7.0. Vortex to mix and add 115 μl of 100% ethanol. Mix by gentle vortexing and freeze in dry ice for 30 minutes. Under these conditions of limiting precipitation with EDTA, the DNA fragments precipitate leaving the unincorporated ^{32}P-labelled nucleotide in solution.

20. Spin down the DNA precipitate in a microcentrifuge for 15 minutes.

21. Repeat steps 6—8.

22. Redissolve the DNA in distilled water as required; usually about 0.5 ml is convenient.

23. Take 5 μl and count on the ^{32}P channel of a liquid scintillation counter.

24. For use in Southern blotting, a nice pattern of molecular weight markers is obtained on a weekend (2½ days) exposure of X-ray film when 5000 cpm of HindIII-λ or HinfI-pAT are loaded on the agarose gel, and when 20 000 cpm of HpaII-pAT fragments are loaded.

Postscript — The labelled polynucleotide markers can be separated from unincorporated ^{32}P-nucleotide by passage down a G50 Sephadex column instead of by limiting ethanol precipitation, if desired. In this case, replace steps 16—22 by step 13, protocol E, Section 3.3.

3.5 REFERENCES

Arrand, J.E. (1985) In: *Nucleic Acid Hybridisation*. Eds. Hames, B.D. and Higgins, S.J., pp. 17—45. IRL Press, Oxford.

Maniatis, T., Fritsch, E.F. and Sambrook, J. (1982) In: *Molecular Cloning: A Laboratory Manual*. Cold Spring Harbor Laboratory, Cold Spring Harbor, NY.

Rigby, P.W.J., Dieckmann, M., Rhodes, C. and Berg, P. (1977) *J. Mol. Biol.* **113**: 237—251.

Southern, E.M. (1975) *J. Mol. Biol.* **98**: 503—517.

Chapter 4

Preparation of RNA from tissues or cells in culture

4.1   PRINCIPLES INVOLVED

Each mammalian cell contains about 10 pg of RNA, of which 80–85% is ribosomal RNA (rRNA) (28S, 18S and 5S varieties), 10–15% is made up of smaller species, such as transfer RNA (tRNA), and 1–5% is messenger RNA (mRNA). The rRNA and tRNA are of defined size and can be banded cleanly by gel electrophoresis or density gradient centrifugation. However, mRNA is very heterogeneous in size, varying from a few hundred to several thousand bases in length, and thus cannot be banded.

RNA can be prepared either from the cytoplasm of cells or from both cytoplasm and nucleus (whole-cell RNA). Cytoplasmic RNA consists mainly of rRNA, tRNA and fully processed mRNA. If the larger heteronuclear mRNA species are to be studied, then RNA will need to be made from the whole cell, including nucleus as well as cytoplasm.

Preparation of cytoplasmic RNA (Favaloro et al., 1980) begins with harvesting a cell pellet, resuspending in buffer and lysing the cell membranes with the detergent NP40. Gentle lysis of the cells breaks the membrane of the cell but not of the nucleus and the nuclei can then be pelleted intact by centrifugation. The detergent SDS is added to the cytoplasmic fraction to inhibit nuclease action and this is then extracted with phenol. Phenol and water are immiscible and protein is extracted into the phenol layer leaving RNA in the aqueous phase. By the addition of two volumes of ethanol to the aqueous phase, the RNA is then precipitated and can be recovered by centrifugation (Fig. 4.1).

Whole-cell RNA can be made by at least two methods – the guanidinium/caesium chloride method (Glisin et al., 1974; Ullrich et al., 1977) or the lithium chloride/urea method (Auffray and Rougeon, 1980). The former procedure is low on labour but involves the use of an ultracentrifuge overnight. The latter method is more labour intensive, requires the use of a sonicator, but requires only low-speed centrifugation. Both methods are described in detail so that RNA can be made according to the facilities available. For the guanidinium/caesium chloride method, cells in culture are harvested and pelleted. The cell pellet is then dispersed in 6 M guanidinium isothiocyanate containing detergent to lyse the cells and a denaturing agent. Caesium chloride is then added to make the solution 2.4 M. This is layered over a cushion of 5.7 M caesium chloride and spun overnight in an ultracentrifuge. This procedure takes advantage of the fact that the buoyant density of RNA in caesium chloride is much greater than that of other cellular macromolecules. During centrifugation, the RNA forms a pellet on the base of the tube while the DNA and protein remain in the upper caesium chloride solution. The RNA pellet is redissolved in a buffer solution and traces of remaining chemicals removed by extraction with chloroform/butanol. The RNA is then cleanly precipitated in ethanol (Fig. 4.2).

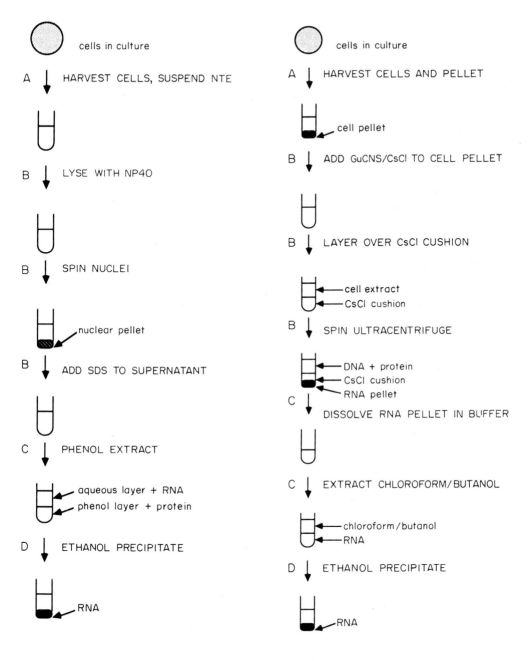

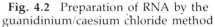

Fig. 4.1 Preparation of cytoplasmic RNA

Fig. 4.2 Preparation of RNA by the guanidinium/caesium chloride method

The lithium chloride/urea method begins with direct lysis of the cells *in situ* on the culture dish in a solution of 3 M lithium chloride/6 M urea/0.1% SDS. The lysate is transferred to a glass tube and sonicated. The sonication breaks down high-molecular-weight DNA while leaving the small RNA molecules intact. If the solution is then left at 4°C overnight, the RNA precipitates out and can be recovered by centrifugation. The RNA is washed with 4 M lithium chloride/8 M urea to remove traces of SDS and is dissolved in buffer solution. It is necessary to then remove protein by phenol extraction. Finally the RNA can be cleanly recovered by ethanol precipitation (Fig. 4.3).

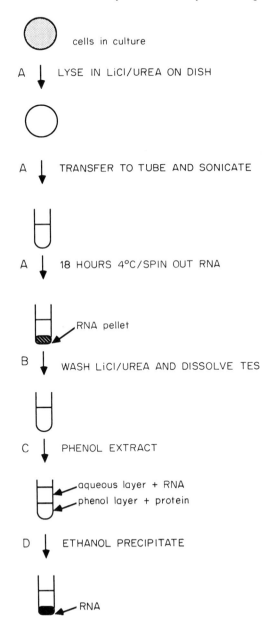

Fig. 4.3 Preparation of RNA by the lithium chloride/urea method

These methods yield a mixture of the cell RNAs. However, mRNA can be isolated alone if required. Nearly all mammalian mRNAs carry a poly(A) sequence at their 3′ ends and this enables them to be purified by affinity chromatography on oligo (dT) cellulose.

The key to preparing good RNA is to minimise ribonuclease activity at all stages. This is done by using both inhibitors of ribonucleases during preparation and by avoiding the introduction of any trace amounts of ribonuclease from fingertips, glassware or solutions. Gloves should, therefore, be worn at all times. All pipettes, glassware and solutions should be autoclaved before use.

4.2 CYTOPLASMIC RNA FROM CULTURED CELLS: WEEKLY SCHEDULE

MONDAY   Steps A−D (see Fig. 4.1)
TUESDAY   RNA ready to use

4.3 CYTOPLASMIC RNA FROM CULTURED CELLS: LABORATORY METHODS

To avoid nuclease contamination, gloves should be worn throughout. All pipettes, glassware and solutions should be autoclaved before use. RNA is not as stable as other macromolecules and attention must be paid to keeping all solutions ice-cold where stated in the protocol. All quantities given are for the preparation of RNA from 10^7 cells and should be scaled up as appropriate.

A. Preparation of cell pellet

RNA can be made from cells growing in monolayer or in suspension culture, and in general, 150 μg RNA will be obtained from 10^7 cells. Cells growing in monolayer can be harvested either with enzyme treatment (e.g. trypsin or collagenase) or by scraping the cells gently off the dish with a rubber policeman. The latter treatment is easier and used for most purposes but the former treatment may be better for delicate primary cultures of cells. Cells growing in suspension, in liquid or semisolid medium, can be transferred to a centrifuge tube simply with a pipette. During harvesting, all cell preparations should be kept ice-cold.

Solutions and chemicals required:
  Phosphate-buffered saline (PBS)

Equipment recommended:
  Pipettes (autoclaved)
  Rubber policeman (autoclaved)
  Sterile plastic centrifuge tubes (Falcon)
  Bucket of ice
  Refrigerated centrifuge at 4°C

PROTOCOL A

1. Cells growing in monolayer culture are washed twice on the dish with PBS. They are then scraped off the dish in cold PBS with a rubber policeman, transferred to a sterile plastic centrifuge tube on ice and pelleted by centrifugation at 4°C at 1000 g for 10 minutes.
2. Cells growing in suspension are transferred directly in cold PBS to a sterile plastic centrifuge tube on ice. The cells are pelleted by centrifugation at 4°C at 1000 g for 10 minutes and the cell pellet washed twice with ice-cold PBS.
3. The cell pellet should be used at once without freezing.

B. Preparation of cytoplasmic fraction

The cell pellet is suspended in NTE buffer and cells are lysed with the detergent NP40. The nuclei are centrifuged out, the upper cytoplasmic fraction collected and the detergent SDS is added to inhibit nuclease action. The solutions are kept ice-cold during all these procedures.

Solutions and chemicals required:
  NTE buffer made as follows:
    1.17 g NaCl
    0.24 g Tris base
    74 mg disodium EDTA
    Dissolve in distilled water
    Bring pH to 7.5 with HCl and make final
      volume up to 200 ml
  NP40
  10% w/v SDS (Serva) in water

Equipment recommended:
  15 ml plastic centrifuge tubes (Falcon)
  Bucket of ice
  Refrigerated centrifuge at 4°C

PROTOCOL B

1. Suspend the cell pellet in 1 ml of NTE buffer by vortexing.
2. Cool on ice for 15 minutes.
3. Make a mixture of 5% NP40 in NTE buffer freshly.
4. Add 0.1 ml of the mixture 5% NP40 in NTE and mix gently.
5. Leave on ice for 10 minutes.
6. Spin at 1000 g for 10 minutes at 4°C.
7. Remove upper cytoplasmic fraction to a clean tube (discard nuclear pellet) and add 0.1 ml 10% SDS solution. Mix by vortexing and keep on ice.

C. Phenol extraction

The cytoplasmic fraction is extracted with phenol. Since phenol and water are immiscible, the protein extracted into the phenol layer is removed from the RNA left in the aqueous layer. During preparation of DNA (Chapter 2) this procedure can be carried out at room temperature but RNA is much less stable and all phenol extraction must be done under ice-cold conditions.

Phenol used in this procedure must be very pure and must be redistilled before use, water saturated and equilibrated with Tris buffer. Nowadays, redistilled phenol can be purchased from companies such as Rathburn Chemicals but any other phenol product must be redistilled before use. The final washes with chloroform alone serve to remove all traces of phenol.

Solutions and chemicals required:
   Chloroform
   Phenol (see below)

Preparation of phenol:
   (i) Take 100 ml water-saturated redistilled phenol (Rathburn Chemicals).
   (ii) Add 100 ml 0.5 M Tris−HCl pH 8.0 containing 1 mM EDTA.
   (iii) Mix well and leave 1 hour at room temperature.
   (iv) Remove aqueous phase and discard.
   (v) Repeat procedures (ii), (iii) and (iv) once more.
   (vi) Add 100 ml 10 mM Tris−HCl pH 8.0 containing 1 mM EDTA.
   (vii) Mix well and leave 1 hour at room temperature.
   (viii) Remove aqueous phase and discard.
   (ix) Repeat procedures (vi), (vii) and (viii) twice more.
   (x) Add 100 ml 10 mM Tris−HCl pH 8.0
   (xi) Store at 4°C in the dark.

With time, phenol will oxidise and change from colourless to an orange colour. This can be slowed down by the addition of 0.02% 8-hydroxyquinoline to the phenol. The solution can, however, be used for 6 months or more.

Phenol causes severe burns and must be handled with care.

Equipment recommended:
  Autoclaved glass pipettes
  Autoclaved Corex glass centrifuge tubes
  Bucket of ice
  Refrigerated Sorvall centrifuge at 4°C with
    rotor SS34 (or equivalent)

PROTOCOL C

 1. Transfer cytoplasmic fraction to an ice-cold Corex glass centrifuge tube placed on ice.
 2. Add an equal volume of phenol : chloroform (1 : 1 by volume) mixture.
 3. Seal tube and mix gently by inverting the tube.
 4. Separate the phases by centrifugation at 7000 g (Sorvall centrifuge with rotor SS34 at 8000 rpm) for 10 minutes at 4°C.
 5. Remove the upper aqueous layer to a clean ice-cold Corex glass centrifuge tube.
 6. Repeat procedures 2–5 once more.
 7. Add an equal volume of chloroform alone.
 8. Seal tube and mix gently by inverting the tube.
 9. Separate the phases by centrifugation as at step 4.
10. Remove the upper aqueous layer to a clean ice-cold Corex glass centrifuge tube.

D. Ethanol precipitation

RNA can be precipitated from an aqueous salt solution by the addition of ethanol.

Solutions and chemicals required:
  100% ethanol (ice-cold)
  70% v/v ethanol in water (ice-cold)
  3 M potassium acetate pH 5.0
  Distilled water

Equipment recommended:
  Autoclaved glass pipettes
  Autoclaved Corex glass centrifuge tubes
  Bucket of ice
  Freezer at −20°C
  Refrigerated Sorvall centrifuge at 4°C with
    rotor SS34 (or equivalent)

PROTOCOL D

1. To the final aqueous phase add 1/10 × volume of 3 M potassium acetate pH 5.0 and mix gently. Keep on ice.
2. Add 2 volumes of ice-cold 100% ethanol and mix gently.
3. Store at −20°C overnight (about 18 hours).
4. Spin down the RNA precipitate in a Sorvall centrifuge (rotor SS34) at 9000 rpm (9700 g) at 4°C for 20 minutes.
5. Pour off the ethanol.
6. Wash the RNA pellet in ice-cold 70% ethanol.
7. Repeat step 4.
8. Pour off the ethanol. Drain and dry the RNA pellet for a few minutes.
9. Redissolve the RNA in sterile distilled water, about 30 μl per 10^7 cells started with. The RNA should dissolve at once and give an RNA solution of about 5 μg/μl, which is optimal for running on an agarose gel and blotting (Chapter 5).

4.4 WHOLE-CELL RNA FROM CULTURED CELLS BY GUANIDINIUM/CAESIUM CHLORIDE METHOD: WEEKLY SCHEDULE

MONDAY	Steps A–B (see Fig. 4.2)
TUESDAY	Steps C–D (see Fig. 4.2)
WEDNESDAY	RNA ready for use

4.5 WHOLE-CELL RNA FROM CULTURED CELLS BY GUANIDINIUM/CAESIUM CHLORIDE METHOD: LABORATORY METHODS

To avoid nuclease contamination, gloves should be worn throughout. All pipettes, glassware and solutions should be autoclaved or filter-sterilised as indicated. Quantities given are for preparation of RNA from 10^7 cells and should be scaled up as appropriate.

A. Preparation of cell pellet

RNA can be made from cells growing in monolayer or in suspension culture, and in general 150 μg RNA will be obtained from 10^7 cells. Cells growing in monolayer can be harvested either with enzyme treatment (e.g. trypsin or collagenase) or by scraping the cells gently off the dish with a rubber policeman. The latter treatment is easier and used for most purposes but the former treatment may be better for delicate primary cultures of cells. Cells growing in suspension, in liquid or semisolid medium, can be transferred to a centrifuge tube simply with a pipette. During harvesting, all cell preparations should be kept ice-cold.

Solutions and chemicals required:
  Phosphate-buffered saline (PBS)

Equipment recommended:
  Pipettes (autoclaved)
  Rubber policeman (autoclaved)
  Sterile plastic centrifuge tubes (Falcon)
  Bucket of ice
  Refrigerated centrifuge at 4°C

PROTOCOL A

1. Cells growing in monolayer culture are washed twice on the dish with PBS. They are then scraped off the dish in cold PBS with a rubber policeman, transferred to a sterile plastic centrifuge tube on ice and pelleted by centrifugation at 4°C at 1000 g for 10 minutes.
2. Cells growing in suspension are transferred directly in cold PBS to a sterile plastic centrifuge tube on ice. The cells are pelleted by centrifugation at 4°C at 1000 g for 10 minutes and the cell pellet washed twice with ice-cold PBS.
3. The cell pellet can either be used at once or stored frozen at −70°C.

B. Purification of RNA by ultracentrifugation

The cell pellet is dispersed in 6 M guanidinium isothiocyanate containing the detergent N-lauroyl-sarcosine to lyse the cells and a denaturing agent, β-mercaptoethanol. The cell extract is made 2.4 M in caesium chloride and layered over a denser 5.7 M caesium chloride solution. During ultracentrifugation, the RNA pellets on the base of the tube leaving DNA and protein in the upper layers. This entire procedure is carried out at room temperature.

Solutions and chemicals required:
  GuCNS solution made as follows:
    70.8 g guanidinium isothiocyanate (Fluka)
    10 ml 50 mM sodium citrate pH 7.0
    (autoclaved)
    0.7 ml β-mercaptoethanol
    0.5 g N-lauroyl sarcosine
    Make volume to 100 ml with distilled water
    Stir mixture continuously for about 1 hour at room temperature until it is fully dissolved
    Filter-sterilise through a 0.45 μm membrane filter (do not autoclave)
    This solution can be kept for up to 1 month at −20°C

  Caesium chloride
  5.7 M CsCl/0.1 M EDTA pH 7.5 made as follows:
    9.6 g caesium chloride
    2 ml 0.5 M EDTA pH 7.5
    8 ml distilled water
    Autoclave the mixture

Equipment recommended:
  Autoclaved glass pipettes
  Beckman SW50.1 5-ml polyallomer tubes
  Balance for weighing
  Beckman ultracentrifuge with rotor SW50.1
    (or equivalent)

PROTOCOL B

1. To the cell pellet add 5 volumes of GuCNS solution at room temperature and disperse by vortexing.
2. Add 1 g of solid caesium chloride to each 2.5 ml of cell homogenate and mix by vortexing.

3. Place 1.2 ml of 5.7 M CsCl/0.1 M EDTA pH 7.5 in a
   Beckman SW50.1 5-ml polyallomer tube and layer up
   to 3.5 ml of the cell homogenate mixture on top.
4. Centrifuge at 35 000 rpm overnight (12−18 hours) at
   15°C in a Beckman ultracentrifuge in a SW50.1 rotor
   (147 000−81 900 g).

C.   Recovery of RNA after ultracentrifugation

The RNA pellet is redissolved in a buffer solution and cleaned by chloroform/butanol
extraction.

Solutions and chemicals required:
TE buffer made as follows:
    100 μl 1 M Tris pH 7.4
    100 μl 0.5 M EDTA
    9.8 ml distilled water
Chloroform
1-butanol

Equipment recommended:
Microcentrifuge tubes 1.5 ml volume (auto-
    claved) (Treff Lab)
Pipette tips (autoclaved)
Bucket of ice
Microcentrifuge at 4°C (Eppendorf or M.S.E.)

PROTOCOL C

1. Remove tube from ultracentrifuge and discard all
   supernatant by pouring it off.
2. Dry the walls of the centrifuge tube thoroughly with
   a piece of clean tissue paper.
3. Dissolve the RNA pellet in 500 μl of TE buffer (this
   is sufficient for up to 2.5 mg of RNA: for approxi-
   mately 2×10^8 cells initially).
4. Transfer to microcentrifuge tube and cool on ice.
5. Add 400 μl of chloroform and 100 μl of 1-butanol.
6. Mix by vortexing.
7. Break the phases by centrifugation for 5 minutes in
   a microcentrifuge at 4°C.
8. Transfer the upper aqueous layer to a clean microcentrifuge tube.

D.   Ethanol precipitation

RNA can then be precipitated from the aqueous solution by addition of salt and ethanol.

Solutions and chemicals required:
100% ethanol (ice-cold)
70% v/v ethanol in water (ice-cold)
3 M potassium acetate pH 5.0
Distilled water

Equipment recommended:
Pipette tips (autoclaved)
Microcentrifuge tubes 1.5 ml volume (auto-
   claved) (Treff Lab)
Bucket of ice
Freezer at $-20°C$
Microcentrifuge at 4°C (Eppendorf or M.S.E.)

PROTOCOL D

1. To the aqueous layer, add $1/10 \times$ volume of 3 M potassium acetate pH 5.0 and mix. Keep on ice.
2. Add 2 volumes of ice-cold 100% ethanol and mix.
3. Store at $-20°C$ overnight (about 18 hours). If the RNA is required urgently, the tube can be placed in dry ice and the RNA is then recoverable after 30 minutes.
4. Spin down the RNA precipitate in a microcentrifuge at 4°C for 15 minutes.
5. Pour off the ethanol.
6. Wash the RNA pellet in ice-cold 70% ethanol.
7. Repeat step 4.
8. Pour off the ethanol. Drain and dry the RNA pellet for a few minutes.
9. Redissolve the RNA in sterile distilled water, about 30 μl per 10^7 cells started with. The RNA should dissolve at once and give an RNA solution of about 5 μg/μl, which is optimal for running on an agarose gel and blotting (Chapter 5).

4.6 WHOLE-CELL RNA FROM CULTURED CELLS BY LITHIUM CHLORIDE/UREA METHOD: WEEKLY SCHEDULE

> MONDAY   Step A (see Fig. 4.3)
> TUESDAY   Steps A–D (see Fig. 4.3)
> WEDNESDAY RNA ready for use

4.7 WHOLE-CELL RNA FROM CULTURED CELLS BY LITHIUM CHLORIDE/UREA METHOD: LABORATORY METHODS

To avoid nuclease contamination, gloves should be worn throughout. All pipettes, glassware and solutions should be autoclaved before use. All quantities given are for preparation of RNA from 10^7 cells and should be scaled up as appropriate.

A. Preparation of cell pellet, lysis and sonication

RNA can be made from cells growing in monolayer or in suspension culture, and in general, 150 μg RNA will be obtained from 10^7 cells. Cells growing in monolayer are washed and lysed *in situ* on the petri dish. Cells growing in suspension, in liquid or semisolid medium, are transferred to a centrifuge tube, pelleted and then lysed. The lysate is then sonicated to break down high-molecular-weight DNA.

> **Solutions and chemicals required:**
>   Phosphate-buffered saline (PBS)
>   4 M LiCl : 8 M urea solution made as follows:
>   42.4 g lithium chloride
>   120 g urea (electrophoresis grade)
>   4.17 ml 3 M sodium acetate pH 5.2 (autoclaved)
>   Make up to 250 ml with autoclaved distilled water
>   Filter-sterilise through a 0.2 μm membrane filter (do not autoclave)
>   Store at 4°C for up to one month
>   3 M LiCl : 6 M urea solution made as follows:
>   30 ml 4 M LiCl : 8 M urea solution (made as above)
>   0.16 ml 3 M sodium acetate pH 5.2 (autoclaved)
>   0.4 ml 10% w/v SDS (Serva) in water
>   Make up to 40 ml with autoclaved distilled water
>   Make freshly as required

Equipment recommended:
  Pipettes (autoclaved)
  Rubber policeman (autoclaved) (for cells in
    monolayer)
  Sterile polypropylene or Corex glass tubes
  Sonicator at 4°C
  Bucket of ice (for cells in suspension)
  Refrigerated centrifuge (for cells in suspension)
  Refrigerator at 4°C

PROTOCOL A

1. Cells growing in monolayer culture are washed once
   on the dish with PBS. 10 ml of 3 M LiCl : 6 M urea
   solution are added to each petri dish containing up
   to 10^7 cells. The cells are scraped off the dish with a
   rubber policeman and transferred to a polypropylene
   or Corex glass tube.
2. Cells growing in suspension are transferred directly
   in cold PBS to a polypropylene or Corex glass tube on
   ice. The cells are pelleted by centrifugation at 4°C at
   1000 g for 10 minutes and the cell pellet washed once
   with ice-cold PBS. 10 ml of 3 M LiCl : 6 M urea solution is added for each 10^7 cells and the cells lysed by
   vortexing.
3. Sonicate at 4°C for four 15-second bursts at high power
   output.
4. Store at 4°C for at least 5 hours, preferably overnight
   for 18 hours.

B.   Preparation of RNA pellet

The RNA is recovered by centrifugation, washed with 4 M lithium chloride/8 M urea solution and dissolved in a Tris/EDTA buffer solution.

Solutions and chemicals required:
   4 M LiCl : 8 M urea solution (for preparation see protocol A)
   10 × TES buffer made as follows:
      30 g Tris base
      50 ml 0.5 M EDTA pH 8.0 (autoclaved)
      2.5 g SDS (Serva)
      Make up to 250 ml with autoclaved distilled water
      Filter-sterilise through a 0.2 μm membrane filter (do not autoclave)
   1 × TES buffer made by dilution of 10 × TES buffer
   Distilled water

Equipment recommended:
   Autoclaved Corex glass tubes
   Refrigerated Sorvall centrifuge at 4°C with
      rotor SS34 (or equivalent)
   Water bath at 65°C

PROTOCOL B

1. Spin down the RNA precipitate in a Sorvall centrifuge (rotor SS34) at 9000 rpm (9700 g) at 4°C for 20 minutes.
2. Pour off aqueous solution and wash the RNA precipitate in 5–10 ml of 4 M LiCl : 8 M urea solution.
3. Repeat step 1.
4. Pour off aqueous solution. Drain and dry the RNA pellet for a few minutes.
5. Dissolve RNA in 1 × TES buffer, using 5 ml for each 10^7 cells at the start. Heat to 65°C in a water bath for 5 minutes, and then cool.

C. Phenol extraction

The RNA is extracted with phenol to remove remaining protein. Since phenol and water are immiscible, the protein extracted into the phenol layer is removed from the RNA left in the aqueous layer.

Phenol used in this procedure must be very pure and must be redistilled before use, water saturated and equilibrated with Tris buffer. Nowadays, redistilled phenol can be purchased from companies, such as Rathburn Chemicals, but any other phenol product must be redistilled before use. Final washes with chloroform alone serve to remove all traces of phenol.

Solutions and chemicals required:
Chloroform
Phenol (see below)

Preparation of phenol:
(i) Take 100 ml water-saturated redistilled phenol (Rathburn Chemicals).
(ii) Add 100 ml 10 × TES buffer (see protocol B).
(iii) Mix well and leave 1 hour at room temperature.
(iv) Remove aqueous phase and discard.
(v) Repeat procedures (ii), (iii) and (iv) once more.
(vi) Add 100 ml 1 × TES buffer (see protocol B).
(vii) Mix well and leave 1 hour at room temperature.
(viii) Remove aqueous phase and discard.
(ix) Repeat procedures (vi), (vii) and (viii) twice more.
(x) Add 100 ml 1 × TES buffer (see protocol B).
(xi) Store at 4°C in the dark.

With time, phenol will oxidise and change from colourless to an orange colour. This can be slowed down by the addition of 0.02% 8-hydroxyquinoline to the phenol. The solution can, however, be used for 6 months or more.

Phenol causes severe burns and must be handled with care.

> **Equipment recommended:**
>   Autoclaved glass pipettes
>   Autoclaved Corex glass centrifuge tubes
>   Bucket of ice
>   Refrigerated Sorvall centrifuge at 4°C with
>   rotor SS34 (or equivalent)

PROTOCOL C

1. Transfer RNA solution to an ice-cold Corex glass centrifuge tube placed on ice.
2. Add an equal volume of phenol : chloroform (1 : 1 by volume) mixture.
3. Seal tube and mix gently by inverting the tube.
4. Separate the phases by centrifugation at 7000 g (Sorvall centrifuge with rotor SS34 at 8000 rpm) for 10 minutes at 4°C.
5. Remove the upper aqueous layer to a clean ice-cold Corex glass centrifuge tube.
6. Repeat procedures 2–5 once more.
7. Add an equal volume of chloroform alone.
8. Seal tube and mix gently by inverting the tube.
9. Separate the phases by centrifugation as at step 4.
10. Remove the upper aqueous layer to a clean ice-cold Corex glass centrifuge tube.

D.   Ethanol precipitation

RNA can be precipitated from an aqueous salt solution by the addition of ethanol.

Solutions and chemicals required:
  100% ethanol (ice-cold)
  70% v/v ethanol in water (ice-cold)
  3 M potassium acetate pH 5.0
  Distilled water

Equipment recommended:
  Autoclaved glass pipettes
  Autoclaved Corex glass centrifuge tubes
  Bucket of ice
  Freezer at $-20°C$
  Refrigerated Sorvall centrifuge at 4°C with
    rotor SS34 (or equivalent)

PROTOCOL D

1. To the final aqueous phase add 1/10 × volume of 3 M potassium acetate pH 5.0 and mix gently. Keep on ice.
2. Add 2 volumes of ice-cold 100% ethanol and mix gently.
3. Store at $-20°C$ overnight (about 18 hours).
4. Spin down the RNA precipitate in a Sorvall centrifuge (rotor SS34) at 9000 rpm (9700 g) at 4°C for 20 minutes.
5. Pour off the ethanol.
6. Wash the RNA pellet in ice-cold 70% ethanol.
7. Repeat step 4.
8. Pour off the ethanol. Drain and dry the RNA pellet for a few minutes.
9. Redissolve the RNA in sterile distilled water, about 30 μl per 10^7 cells started with. The RNA should dissolve at once and give an RNA solution of about 5 $\mu g/\mu l$, which is optimal for running on an agarose gel and blotting (Chapter 5).

4.8 RNA FROM TISSUES

RNA is prepared from tissues in an analogous way to that described for RNA from cultured cells. The main problem encountered here is that of disaggregating the cells in the tissue to produce a single-cell suspension. This can be done simply by grinding the frozen tissue with a pestle and mortar in liquid nitrogen. However, if it is available, use of a dismembranator makes life easier. Since the cells are now severely damaged, cytoplasmic RNA cannot be prepared. However, whole-cell RNA can be made by either the guanidinium/caesium chloride method (section 4.5) or the lithium chloride/urea method (section 4.7).

A. Disaggregation of cells in the tissue

Equipment recommended:
Pestle and mortar or dismembranator
Liquid nitrogen

PROTOCOL A

1. Freeze the tissue in dry ice or liquid nitrogen.
2. Grind the tissue to a fine powder. This can be done either by simply using a pestle and mortar, keeping the tissue frozen by covering in liquid nitrogen while grinding, or by using a dismembranator.

B. Preparation of RNA

RNA can then be made from the tissue powder, either starting at Section 4.5, protocol B, step 1 or starting at Section 4.7, protocol A, step 2.

4.9 CONDITION OF RNA

1. The concentration of the RNA can be measured using optical density:
   Take 1 μl of the RNA solution and add 500 μl of water.
   Read optical density at 260 nm (OD_{260}).
   A solution of RNA in water of 1 mg/ml gives an OD_{260} of 25.
   Thus $\mu g/\mu l$ in RNA solution = $1/25 \times OD_{260} \times 500$.
2. When RNA is prepared, the molecules should be intact and not broken down. This can be ensured for each RNA sample by running a small aliquot on a 1.5% agarose gel. This is done by following protocol A in Chapter 5. About 5 μg of RNA are run on the gel. Bands of 28S, 18S and 5S rRNA should be clearly visible on the stained gel. If RNA is broken down badly, these bands are not visible and the RNA appears as a smear down the gel track.
3. RNA in solution can be stored at $-70°C$ for several weeks and often months. For long-term storage, however, RNA is more stable if stored in ethanol as at stage 3 of protocol D (see any of the preparation methods).

4.10 SELECTION OF POLY(A)⁺ RNA

mRNA can be isolated cleanly from other cellular RNA by affinity chromatography on oligo (dT) cellulose (Edmonds *et al.*, 1971; Aviv and Leder, 1972). The long tail of adenine (A)-containing nucleotides of mRNA binds to a complementary string of thymine (T)-containing nucleotides immobilised in a column of cellulose, while all other RNA molecules wash straight through the column. The isolated poly(A)⁺ RNA can be eluted finally from the column by reducing the salt concentration. This procedure should be carried out under cool conditions. The ideal way is to do everything in a cold room at 4°C. If such a room is not available, it can be done by keeping all solutions ice-cold when applied to the column and collecting all eluates in tubes in a bucket of ice.

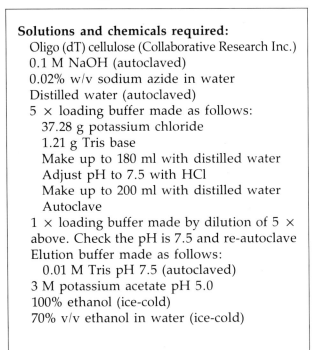

Solutions and chemicals required:
  Oligo (dT) cellulose (Collaborative Research Inc.)
  0.1 M NaOH (autoclaved)
  0.02% w/v sodium azide in water
  Distilled water (autoclaved)
  5 × loading buffer made as follows:
    37.28 g potassium chloride
    1.21 g Tris base
    Make up to 180 ml with distilled water
    Adjust pH to 7.5 with HCl
    Make up to 200 ml with distilled water
    Autoclave
  1 × loading buffer made by dilution of 5 ×
  above. Check the pH is 7.5 and re-autoclave
  Elution buffer made as follows:
    0.01 M Tris pH 7.5 (autoclaved)
  3 M potassium acetate pH 5.0
  100% ethanol (ice-cold)
  70% v/v ethanol in water (ice-cold)

Equipment recommended:
  Sterile disposable column (1 ml volume)
  Pipette tips (autoclaved)
  Microcentrifuge tubes 1.5 ml volume (auto-
    claved) (Treff Lab)
  2 buckets of ice
  Spectrophotometer
  Freezer at −20°C
  Corex glass tubes (autoclaved)
  Refrigerated Sorvall centrifuge at 4°C with
    rotor SS34 (or equivalent)

PROTOCOL

1. Equilibrate 0.5 g of oligo (dT) cellulose in 1 × loading buffer.
2. Pour a small column in a disposable column or a plastic syringe plugged at the base with polyester wool.
3. Wash the column with at least 5 volumes of ice-cold 1 × loading buffer, running at a speed of about 1 drop every 20–30 seconds.
4. Take required amount of RNA in 1 ml of water and add 0.25 ml of 5 × loading buffer.
5. Apply RNA to the column and collect the eluate in a tube on ice.
6. Reapply the eluate to the column five times.
7. Wash the column with about 20 ml ice-cold 1 × loading buffer.
8. Treat the column with 6 ml of ice-cold elution buffer and collect 6 × 1 ml fractions in sterile tubes in an ice bath.
9. Take a 50 μl aliquot of each fraction collected, add to 450 μl water and measure OD_{260}. A peak of mRNA is usually seen in tubes 2–4.
10. Pool the fractions containing the mRNA and place in a Corex glass tube. Add 1/10 × volume of 3 M potassium acetate pH 5.0 and mix gently, keeping ice-cold. Add two volumes of ice-cold 100% ethanol, mix and store at −20°C overnight (18 hours).
11. Spin down the mRNA precipitate in a Sorvall centrifuge (rotor SS34) at 9000 rpm (9700 g) at 4°C for 20 minutes.
12. Pour off the ethanol.
13. Wash the mRNA pellet in ice-cold 70% ethanol.
14. Repeat steps 11 and 12.
15. Drain and dry the mRNA for a few minutes and then redissolve in sterile distilled water.
16. After use, the oligo (dT) cellulose column should be washed with 20 ml 0.1 M NaOH and then 20 ml of 0.02% azide. The column can be stored at 4°C in 0.02% azide.

4.11 REFERENCES

Auffray, C. and Rougeon, F. (1980) *Eur. J. Biochem.* **107**: 303–314.
Aviv, H. and Leder, P. (1972) *Proc. Natl. Acad. Sci. USA* **69**: 1408–1412.
Edmonds, M., Vaughn, Jr., M.H. and Nakazato, H. (1971) *Proc. Natl. Acad. Sci. USA* **68**: 1336–1340.
Favaloro, J., Freisman, R. and Kamen, R. (1980) *Methods Enzymol.* **65**: 718–749.
Glisin, V., Crkvenjakov, R. and Byus, C. (1974) *Biochemistry* **13**: 2633–2637.
Ullrich, A., Shine, J., Chirgwin, J., Pictet, R., Tischer, E., Rutter, W.J. and Goodman, H.M. (1977) *Science* **196**: 1313–1319.

Chapter 5

RNA analysis by Northern blotting

5.1 PRINCIPLES INVOLVED

Northern blotting techniques allow both size estimation and relative quantitations of any RNA species. The principles involved are illustrated in Fig. 5.1. RNA fragments are denatured and separated by size on electrophoresis through agarose gels containing formaldehyde. The single-stranded RNA pieces can then be transferred to a sheet of nitrocellulose. It is this transfer of electrophoretically-resolved RNA fragments to nitrocellulose filters which is known as 'Northern blotting' (Alwine, Kemp and Stark, 1977). This entails laying a sheet of nitrocellulose (which acts as a filter) on top of the gel and establishing a flow of buffer through the gel and the nitrocellulose filter. The buffer carries the RNA fragments upwards from the gel to the nitrocellulose, where they subsequently bind. A single-stranded nucleic acid probe specific for the RNA under study is then radiolabelled and hybridised to the filter. This probe can be a purified RNA, cDNA or cloned fragment of genomic DNA. Whichever is chosen, the labelled probe will hybridise to any RNA fragment on the filter that contains complementary nucleotide sequences. Autoradiography of the nitrocellulose filter will reveal the position of each piece of RNA containing sequences homologous to any part of the probe used.

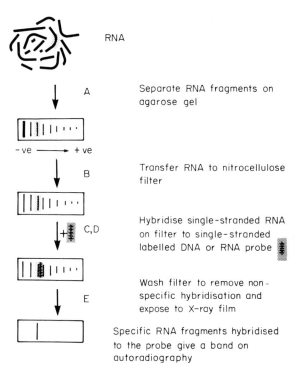

Fig. 5.1 Principles of Northern blotting

The molecular size of each RNA can be estimated by comparing its position relative to the positions of the 28S, 18S and 5S rRNA bands visible on the agarose gel when stained with ethidium bromide. When fully denatured, the rate of migration of each RNA through the gel is proportional to the $\log_{10}$ of its molecular weight. A conversion graph is given in Fig. 5.2 for converting from S-value of RNA to its molecular size in nucleotides.

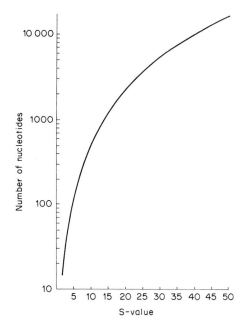

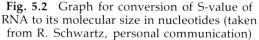

Fig. 5.2 Graph for conversion of S-value of RNA to its molecular size in nucleotides (taken from R. Schwartz, personal communication)

5.2 WEEKLY SCHEDULE

MONDAY	Run RNA gel for 3 hours (protocol A)
	Stain gel and photograph (protocol B, 1–4)
	Blot RNA onto nitrocellulose overnight (protocol B, 5)
TUESDAY	Disassemble blot and bake (protocol B, 6)
	Nick translate DNA to make ^{32}P-probe (protocol C)
	Hybridise blot to probe overnight (protocol D)
WEDNESDAY	Wash blot, dry blot, and put onto autoradiography (protocol E)

5.3 LABORATORY METHODS

Nuclease contamination must be avoided in procedure C: for that protocol gloves must be worn throughout. Pipettes, glassware and solutions should all be autoclaved before use. These precautions are not necessary for the other protocols.

A. Prepare and run an agarose gel for RNA

RNA molecules are separated by size on agarose gel electrophoresis.

Solutions and chemicals required:
  10 × MOPS buffer made as follows:
    41.8 g MOPS (Sigma)
    3.72 g EDTA (di Na)
    4.10 g sodium acetate
    Make volume to 900 ml with water
    Adjust pH to 7.0 (with *very* strong NaOH)
    Make volume to 1 litre with water
  Agarose powder (SEAKEM-ME) (FMC Corp.)
  Distilled water
  Sample buffer made as follows:
    200 μl 10 × MOPS buffer
    1.0 ml formamide
    356 μl formaldehyde (38% stock)
  Dye solution made as follows:
    15% Ficoll 400 (w/v)
    4% bromophenol blue (w/v)
    In distilled water
  Formaldehyde (38% stock)

Equipment recommended:
  Microwave oven or boiling water bath
  Agarose gel electrophoresis equipment commercially available or home-made)
  Electrophoresis power supply
  Water bath at 60°C

PROTOCOL A

1. Prepare mould for the gel.
   Many systems are available commercially for pouring gels but a simple home-made system is illustrated in Fig. 5.3.
   A perspex plate is levelled using a spirit level.
   A glass plate (14 cm × 19 cm) is used as the mould. Its edges are sealed with PVC electrical tape (or radioactive/biohazard-labelled tape) so that it forms a wall 1 cm high all around the plate. This is placed on the level perspex plate. A simple perspex comb is assembled at one end of the plate so that there is about 1 mm of space between the base of the teeth and the glass plate.
2. Weigh 2.25 g of agarose into a 200 ml conical flask.
   Add 15 ml of 10 × MOPS buffer.
   Add 108 ml of water.
   Heat in microwave oven or boiling water bath until agarose dissolves.
3. Add 27 ml of formaldehyde (38% stock).
   Pour agarose onto gel mould and allow to set for 30 minutes.
   NB The recipe given here provides a 1.5% agarose gel, which is useful for most purposes. To separate very large or very small RNA fragments, the concentration of agarose can be varied.
   NBB Formaldehyde should not be inhaled. These gels should be poured in a fume hood.

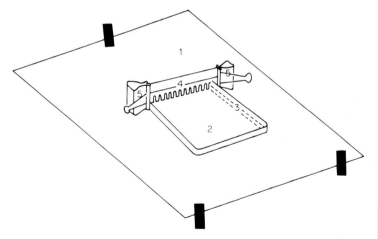

Fig. 5.3 Mould for pouring agarose gels. 1 = perspex plate levelled with three screw feet; 2 = glass plate 14 cm × 19 cm; 3 = PVC tape; 4 = perspex comb to mould wells in the gel; 5 = clamp to hold comb in position

4. The gel tank is prepared and connected to an electro-phoresis power supply. Many systems are available commercially but a simple home-made system is illustrated in Fig. 5.4.
5. Place 1600 ml of 1 × MOPS buffer into the gel tank.
6. Put the gel into the tank.
   The surface of the gel should be just submerged (by about 1 cm depth of buffer).
7. Take required amount of RNA in 9 μl water. (Up to 70 μg of RNA can be loaded per well in the gel.) Add 31 μl sample buffer.
   Heat to 60°C for 5 minutes (to denature RNA).
8. Add 8 μl dye solution at once and load quickly into the wells of the gel (before RNA renatures).
9. Run gel at 150 V constant voltage or 90 mA constant current for 3 hours, or until the blue dye has run to 100–150 mm from the well. All samples are run from the negative to the positive electrode.

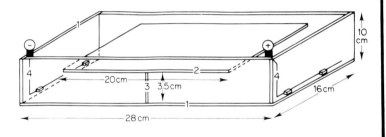

Fig. 5.4 Agarose gel electrophoresis system. A perspex box (1) made of 0.6 cm thick perspex, with a lid, has an internal perspex shelf (2) moulded to the sides of the box. The box is divided into two compartments with a perspex divider (3). Two electrodes are made of platinum wire (4), one in each chamber

B. Northern transfers (blotting)

This procedure was first described by Alwine, Kemp and Stark (1977) and is an adaptation of the method for blotting DNA described by Southern (1975). It involves transfer of the electrophoretically resolved RNA fragments from the agarose gel onto a nitrocellulose filter.

Solutions and chemicals required:
10 mg/ml ethidium bromide (BDH) in water
  (store in the dark)
1 × MOPS buffer (diluted from 10 × MOPS
buffer as in protocol A)
Distilled water
20 × SSC solution made as follows:
  174 g sodium chloride
  88.2 g trisodium citrate dihydrate
  Dissolve in 800 ml distilled water
  Adjust pH to 7.0 with 5 M NaOH
  Make volume up to 1 litre with water
Nitrocellulose. This is available from many
commercial sources (recommended: Sartorius
filters type SM26, pore size 0.1 μm).

Equipment recommended:
Ultraviolet light
2 plastic trays (25 cm × 30 cm × 5 cm high)
2 large rubber bungs (size 49)
2 glass plates (14 cm × 19 cm)
Whatman filter paper (No. 1 and No. 3mm)
One packet of paper hand towels
1 kilogram weight (e.g. a 500 ml bottle filled
with water)
Clingfilm
Vacuum oven

PROTOCOL B

1. Place the gel in a plastic tray of convenient dimensions.
2. RNA in the gel is stained with 4 drops of ethidium bromide (10 μg/ml in water) in 300 ml of 1 × MOPS buffer for 30 minutes.
3. Destain the gel by washing in 300 ml of 1 × MOPS buffer for 30 minutes.
4. RNA can now be visualised under ultraviolet light. Three bands of 28S, 18S and 5S rRNA are evident, and their positions measured from the well in the gel. A photograph can be taken for a permanent record.

5. The blot is set up as shown in Fig. 5.5. This is done in an analogous way to Southern blotting except that 20 × SSC is used instead of 10 × SSC.
   (a) Wet the nitrocellulose filter in distilled water in a plastic tray, then transfer to 20 × SSC solution.
   (b) In another plastic tray, place two large rubber bungs on the bottom.
   (c) Place a glass plate (14 cm × 19 cm) on top.
   (d) Now place a sheet of 3mm Whatman paper over the glass plate so that it acts as wicks on either side of the plate.
   (e) Add 600 ml of 20 × SSC solution into the tray and soak the Whatman paper, making sure that all air bubbles are removed.
   (f) Carefully place the gel on top of the Whatman paper, again removing all air bubbles.
   (g) Now cover all edges of the gel using four pieces of clingfilm, tucking excess under the tray but taking care not to cover any areas of the gel containing RNA. This ensures that the 20 × SSC solution passes evenly through the gel.
   (h) Lay the wetted nitrocellulose sheet onto the gel surface, taking care to exclude air bubbles.
   (i) Lay three sheets of presoaked (20 × SSC) Whatman No. 1 paper over the filter.
   (j) Place about 8 cm of dry Kleenex Hi-Dri paper towels over the Whatman No. 1 paper and place a glass plate on top of this. Now put a kilogram weight on top and leave to blot overnight (a minimum of 24 hours). RNA takes rather longer to transfer than DNA and if large RNA molecules of >28S are to be studied, it is advisable to put on new dry Kleenex Hi-Dri paper towels after 24 hours and leave the blotting for a second 24-hour period.

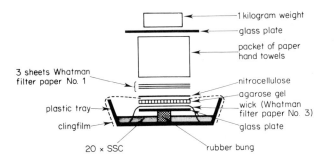

Fig. 5.5 Diagram of apparatus for the Northern blot

6. Disassemble blot the next morning.
   Place the filter (TAKE CARE: with DNA upwards!) on dry tissues.
   Leave to dry in air for about 30 minutes.
   Check the gel under ultraviolet light to ensure that all the RNA has transferred.
   Place the dried filter in a small envelope of filter paper (Whatman No. 3mm) and bake in a vacuum oven at 80°C for 3–5 hours.

C. To make ^{32}P-labelled probe by nick translation of DNA

Radiolabelled probes for nucleic acid hybridisation can be prepared by a variety of methods, which are discussed in more detail elsewhere (Arrand, 1985). In this protocol, one reliable method only is presented. This involves labelling the probe by a nick translation reaction (Rigby et al., 1977). The enzyme DNase I is used to create single-strand nicks in double-stranded DNA. Then the $5' \rightarrow 3'$ exonuclease and $5' \rightarrow 3'$ polymerase actions of E. coli DNA polymerase I are used to incorporate radiolabelled deoxyribonucleotides while repairing the nicks. Finally, the DNA is denatured with sodium hydroxide and free unincorporated deoxyribonucleotides are separated from the DNA by column chromatography. In practice, the end result of hybridisation is usually cleaner if DNA used in this reaction is a piece of DNA under 2000 bases in length and separated from any plasmid vector used for cloning (see Chapter 6).

Equipment recommended:
^{32}P-dCTP emits beta-radiation and should be handled with care.
All work should be done behind perspex screens (1 cm thick) to protect the operator from radiation. Usual precautions for handling ^{32}P-labelled chemicals must be observed at all times.
Geiger counter (for beta-emission)
Microcentrifuge tubes 1.5 ml volume (autoclaved) (Treff Lab)
Water bath at 14°C (use a water bath in a cold room)
5 ml sterile disposable plastic pipette
Siliconised sterile glass wool
Sterile tubing to set up a chromatography column
Scintillation counting facilities

Solutions and chemicals required:

Distilled water (autoclaved)

10 × nick translation (10 × NT) buffer, made as follows:

    500 μl 1 M Tris—HCl pH 8.0 (autoclaved)

    100 μl 0.5 M magnesium chloride (autoclaved)

    10 μl β-mercaptoethanol

    390 μl distilled water (autoclaved)

20 μM dATP, dGTP and dTTP, made as follows:
Each deoxyribonucleotide triphosphate is dissolved in sterile distilled water to give a concentration of 10 mM. The pH of each solution is then adjusted to pH 7.0 using a solution of 0.05 M Tris base. The concentration of each solution is checked by reading the optical density of a small aliquot diluted in water (the molar extinction coefficient for the bases is:

    A 1.54×10^4 $M^{-1}cm^{-1}$ at 259 nm;

    G 1.37×10^4 $M^{-1}cm^{-1}$ at 253 nm;

    T 7.4×10^3 $M^{-1}cm^{-1}$ at 260 nm).

These 10 mM stock solutions can be stored at $-20°$C and diluted in water for use at 20 μM concentration.

^{32}P-dCTP purchased from Amersham (PB10205)

DNA probe (prepared as described in Chapter 6)

1 mg/ml bovine serum albumin (BSA) (Sigma, Fraction V powder) in autoclaved distilled water

DNase I (Worthington). A stock solution of 100 μg/ml can be stored at $-20°$C. This should be diluted to 100 ng/ml for use but at this lower concentration continual thawing and refreezing is not recommended. Aliquots should be stored frozen and thawed once only before discarding.

DNA polymerase I (Boehringer)

5 M sodium hydroxide (autoclaved)

G50 column buffer, made as follows:

    10 ml 5 M sodium chloride (autoclaved)

    5 ml 1 M Tris—HCl pH 7.4 (autoclaved)

    5 ml 10% w/v SDS (Serva) in water (filter-sterilised)

    480 ml distilled water (autoclaved)

Sephadex G50 fine (Pharmacia) preswollen in the G50 column buffer

PROTOCOL C

Prepare nick translation mix in a 1.5-ml microcentrifuge tube (this should be done behind perspex screens)
1. 5 μl dATP (20 μM).
2. 5 μl dTTP (20 μM).
3. 5 μl dGTP (20 μM).
4. 5 μl ^{32}P-dCTP (50 μCi/reaction).
5. 0.5 μg DNA (volume 0.5 μl).
6. 24.5 μl water.
7. 5 μl 10 × NT buffer.
8. Add 1 μl of 1 mg/ml BSA to stabilise the DNase which is added later.
9. Add 2.5 μl of DNase I (100 ng/ml); leave at room temperature for 3 minutes.
10. Add 1 μl of DNA polymerase I.
11. Leave at 14°C for 2 hours.
12. Add 3.3 μl of 5 M NaOH to the mix, and leave at room temperature for 10 minutes.
13. Separate ^{32}P-labelled DNA from ^{32}P-dCTP on a column of Sephadex G50 fine.
    (a) Prepare a column of Sephadex G50 fine in a disposable 5 ml pipette plugged with siliconised sterile glass wool. All column equipment should be sterile. Equilibrate the column with column buffer.
    (b) Apply the reaction mix to the column and elute with column buffer. Monitor effluent with a Geiger counter.
    (c) Two peaks are formed on the column. The first peak is the labelled DNA. The second peak is ^{32}P-dCTP. The first peak is collected in a volume of 1–2 ml.
    (d) Count a 5 μl aliquot by liquid scintillation counting to estimate the specific activity (cpm/μg) of the DNA.

D. Hybridisation

Single-stranded RNA immobilised on the nitrocellulose filter is hybridised to the single-stranded radiolabelled DNA probe. Several methods are available for this reaction, but, commonly, formamide or dextran sulphate are used. Dextran sulphate, although viscous to handle, provides a reliable and sensitive assay method and is the method of choice described here. The procedure described here provides high stringency and only closely homologous RNA-DNA sequences will hybridise. If the DNA probe is unlikely to be closely related to the RNA under examination, the stringency can be lowered by either lowering the temperature or increasing the salt concentration (i.e. amount of 20 × SSC solution added).

Solutions and chemicals required:
Distilled water
20 × SSC solution made as follows:
  174 g sodium chloride
  88.2 g trisodium citrate dihydrate
  Dissolve in 800 ml distilled water
  Adjust pH to 7.0 with 5 M NaOH
  Make volume up to 1 litre with water
25 × Denhardt's solution, made as follows:
  2 g bovine serum albumin (BSA) (Sigma, Fraction V)
  2 g Ficoll 400 (Pharmacia)
  2 g polyvinyl pyrrolidone (Sigma)
  Make up to 400 ml with water
  Filter-sterilise by passing through a 0.45 μm filter
10% w/v SDS (Serva) in water. Sodium dodecylsulphate (Serva) is dissolved in water (10 g/100 ml) and filter-sterilised (0.2 μm).
Single-stranded (SS) DNA solution, made as follows:
  1 mg/ml calf thymus DNA (Sigma); dissolve in boiling water
  Sonicate (amplitude low, 1 minute), then boil again
  Store in a sterile bottle at 4°C
50% dextran sulphate solution, made as follows:
  500 g (Pharmacia sodium salt), plus 800 ml water
  Shake occasionally by hand over several days to dissolve
  Make volume to 1 litre with water

> **Equipment recommended:**
> Plastic bags and a plastic-bag sealer
> (any commercially available system sold for
> domestic use is suitable)
> An incubator set at 65°C with shaking facilities
> (if this is not available, bags can be submerged
> in a shaking water bath)

PROTOCOL D

1. Prepare solution A:
   - 3 ml 20 × SSC solution
   - 8 ml 25 × Denhardt's solution
   - 9 ml distilled water

   Place the nitrocellulose blot inside a plastic bag. Add solution A. Expel air bubbles. Seal bag. Incubate for 1 hour at 65°C with vigorous shaking.

2. Prepare solution B:
   - 1 ml 20 × SSC solution
   - 8 ml 25 × Denhardt's solution
   - 0.2 ml 10% SDS solution
   - 1 ml SS DNA 1 mg/ml
   - 10 ml distilled water

   Cut off the corner of the plastic bag. Pour out solution A. Add solution B. Reseal. Incubate for 1 hour at 65°C with vigorous shaking.

3. Prepare solution C:
   - 1 ml 20 × SSC solution
   - 8 ml 25 × Denhardt's solution
   - 0.2 ml 10% SDS solution
   - 1 ml SS DNA 1 mg/ml
   - 6 ml distilled water
   - 4 ml 50% dextran sulphate

   Cut off the corner of the plastic bag. Pour out solution B. Add solution C. Reseal. Incubate for 1 hour at 65°C with vigorous shaking.

4. Prepare solution D:

   This is identical to solution C but contains in addition 5–10 million cpm of ^{32}P-labelled DNA probe.

   Cut off the corner of the plastic bag. Pour out solution C. Add solution D. Reseal. Incubate for 15–18 hours at 65°C with vigorous shaking.

E.   Washing the blot

After hybridisation, the Northern blot is washed to remove nonspecific hybridisation. The important step in this procedure is step 3, which determines the stringency required. The conditions described here provide a high stringency wash. If the DNA probe is not closely related to the RNA under examination, the stringency of the wash in step 3 should be lowered by either decreasing the temperature or increasing the salt concentration (i.e. amount of 20 × SSC added).

Solutions and chemicals required:
  20 × SSC solution made as follows:
    174 g sodium chloride
    88.2 g trisodium citrate dihydrate
    Dissolve in 800 ml distilled water
    Adjust pH to 7.0 with 5 M NaOH
    Make volume up to 1 litre with water
  25 × Denhardt's solution, made as follows:
    2 g bovine serum albumin (Sigma)
    2 g Ficoll 400 (Pharmacia)
    2 g polyvinylpyrrolidone (Sigma)
    Make up to 400 ml with water
    Filter-sterilise at 0.45 μm
10% w/v SDS (Serva) in water and filter-sterilised at 0.2 μm
Distilled water

Equipment recommended:
  An incubator set at 65°C with shaking facilities (if this is not available, washing can be done in a shaking water bath)
  Plastic tray
  Clingfilm
  Whatman filter paper (No. 3mm)
  X-ray film cassette
  X-ray film (recommended: Kodak XAR5)
  Facilities for developing X-ray film

PROTOCOL E

1. Remove the nitrocellulose filter from the plastic bag in which hybridisation has taken place and put the filter into a plastic tray.
2. Wash at 65°C with 2 × 10-minute followed by 2 × 30-minute washes, using 100 ml each time of solution made as indicated below:
   - 20 ml 20 × SSC solution
   - 160 ml 25 × Denhardt's solution
   - 4 ml 10% SDS solution
   - 216 ml distilled water
3. Wash at 65°C with 2 × 30-minute washes of 100 ml each of solution made as indicated below:
   - 2 ml 20 × SSC solution
   - 2 ml 10% SDS solution
   - 196 ml distilled water
4. Wash the blot with 4−5 rinses at room temperature (1 minute each) with 100 ml each of solution made as indicated below:
   - 75 ml 20 × SSC solution
   - 425 ml distilled water
5. While the blot is still wet, it should be scanned with a Geiger counter. If a large amount of radioactivity is detected scattered all over the filter, there may be nonspecifically hybridised DNA still bound and the blot may need to be washed for longer (repeat steps 3 and 4).
6. After washing, allow the blot to air dry. Then place onto 3mm Whatman paper and cover in clingfilm to prevent the X-ray film sticking to the blot. Expose to X-ray film.

5.4 REFERENCES

Alwine, J.C., Kemp, D.J. and Stark, G.R. (1977) *Proc. Natl. Acad. Sci. USA* **74:** 5350−5354.

Arrand, J.E. (1985) In: *Nucleic Acid Hybridisation*. Eds. Hames, B.D. and Higgins, S.J., pp. 17−45. IRL Press, Oxford.

Rigby, P.W.J., Dieckmann, M., Rhodes, C. and Berg, P. (1977) *J. Mol. Biol.* **113:** 237−251.

Southern, E.M. (1975) *J. Mol. Biol.* **98:** 503−517.

Chapter 6

Preparation of DNA for probes

6.1   PRINCIPLES INVOLVED

Most DNA sequences used as probes for Southern and Northern blotting will have been isolated and cloned into a plasmid vector. No attempt is made here to discuss any of the strategies for gene cloning. However, it is necessary to prepare large quantities of the DNA for use in the nick translation reaction and then as a probe, and it is these methods which are described here.

Plasmids are extrachromosomal genetic elements found in a variety of bacterial strains. They are double-stranded closed circular DNA molecules, ranging in size from 1 kb to 200 kb, and they contain genes coding for enzymes that under certain conditions are advantageous to the bacterial host. The most common phenotype used in cloning is that of resistance to antibiotics. The cloned gene will have been inserted into the plasmid at a specific restriction-enzyme site (which must be known). General features of the plasmid structure are given in Fig. 6.1.

Plasmid DNA is amplified in the bacteria. Thus the first step is to put the plasmid DNA into bacterial cells, a process known as transformation. The bacteria are treated with calcium chloride, which makes some of the cells temporarily permeable to small DNA molecules (Mandel and Higa, 1970; Cohen, Chang and Hsu, 1973). Use of the new phenotype (usually resistance to the antibiotics ampicillin or tetracycline) conferred on the recipients by the plasmid allows simple selection of the bacteria which have been successfully transformed.

The bacteria are then left to grow for some hours, during which time the plasmids also replicate within the cells, giving many copies per cell. The bacteria are then harvested and lysed, and plasmid DNA is isolated from the bacterial host cell DNA. Initially, a differential precipitation step is performed in which the large strands of bacterial genomic DNA, entangled in the remnants of lysed cells, are preferentially removed from the lysate. Final purification of the plasmid is on caesium chloride gradients. The DNA is centrifuged to equilibrium in the gradient in the presence of saturating quantities of the intercalating dye, ethidium bromide. Plasmid DNA, which is covalently closed circular DNA, binds much less of the dye than the linear bacterial genomic DNA and hence bands at a higher density in the caesium chloride gradient (Radloff et al., 1967).

The purified plasmid is then ready for use in a nick translation reaction (Chapter 3, protocol E; Chapter 5, protocol C). In practice, however, final autoradiographs of Southern (Chapter 3) or Northern (Chapter 5) blots have lower backgrounds if the radiolabelled probe contains only the cloned sequences of interest and no plasmid sequences. Thus, finally, the plasmid is cut with the appropriate restriction enzyme used for the cloning originally. This releases the cloned gene from the plasmid and the two DNAs can be separated by agarose gel electrophoresis. The principles involved in these methods are illustrated in Fig. 6.2.

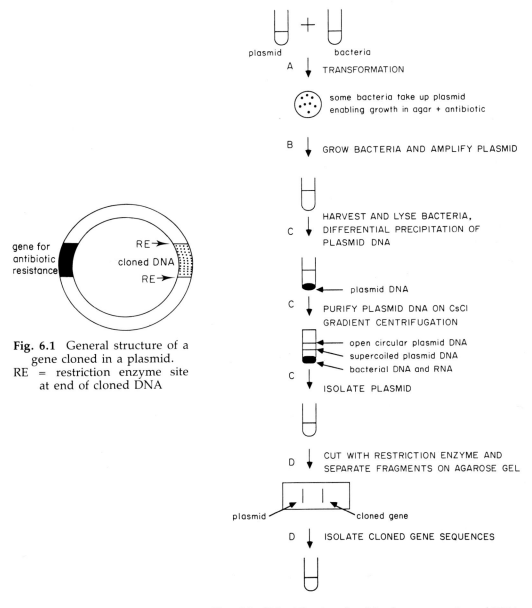

Fig. 6.1 General structure of a
gene cloned in a plasmid.
RE = restriction enzyme site
at end of cloned DNA

Fig. 6.2 Principles involved in the preparation of DNA
for probes used in Southern and Northern blotting

6.2 WEEKLY SCHEDULE

MONDAY	Prepare agar dishes for bacterial growth (protocol A, 1)
	Grow stock culture of bacteria overnight (protocol A, 2)
TUESDAY	Transform bacteria (protocol A, 3–4)
	Grow bacteria on agar dishes overnight (protocol A, 5)
WEDNESDAY	Select bacterial clone and amplify in a small culture overnight (protocol B, 1–4)
THURSDAY	Amplify bacteria in large cultures overnight (protocol B, 5–8)
FRIDAY	Harvest and lyse bacteria protocol C, 1–6)
	Differential precipitation of plasmid DNA (protocol C, 7–15)
	Caesium chloride gradient centrifugation over weekend (protocol C, 16–22)
MONDAY	Prepare plasmid DNA from gradient (protocol C, 23–32)
	Restriction-enzyme digestion overnight (protocol D, 1)
TUESDAY	Isolate cloned gene from plasmid on agarose gel (protocol D, 2–3)
	Ethanol precipitate overnight (protocol D, 4)
WEDNESDAY	DNA probe ready for use in nick translation reaction (protocol D, 4)

6.3 LABORATORY METHODS

For all these methods, pipettes, glassware and solutions must be autoclaved before use. Gloves must be worn throughout. For protocols A and B, this must extend to use of sterile culture technique and all manipulations must be carried out either in a laminar flow culture hood or using sterile flaming techniques for bacteriological culture work. For protocols C and D, use of autoclaving is necessary only to prevent nuclease contamination and sterile technique is not necessary.

A. Transformation of bacteria with plasmid

The bacteria are treated with calcium chloride, which makes some of them temporarily permeable to plasmid molecules. The bacteria are then grown overnight in agar containing either ampicillin or tetracycline. Using the gene for antibiotic resistance contained in the plasmid, simple selection is achieved of the bacteria that have been transformed. It is necessary to know for each plasmid which antibiotic resistance is operative.

Solutions and chemicals required:
Bacto-agar
L-broth made as follows:
    10 g Bacto-tryptone
    5 g Bacto-yeast extract
    10 g sodium chloride
    Make volume to 1 litre with distilled water
    Adjust pH to 7.5 if necessary with sodium hydroxide
    Autoclave
Appropriate antibiotic
    — ampicillin 50 mg/ml in water (filter-sterilised through 0.2-μm filter)
    — tetracycline 15 mg/ml in methanol (filter-sterilised through 0.2-μm filter)
2 M calcium chloride (filter-sterilised through 0.2-μm filter)
Distilled water (autoclaved)
TEN buffer made as follows:
    1 ml Tris–HCl pH 7.4
    0.2 ml 0.5 M EDTA
    4 ml 5 M sodium chloride
    Make up to 100 ml with distilled water
    Filter-sterilise through 0.2-μm filter
Bacterial cells strain HB101
1 μg of plasmid DNA

Equipment recommended:
 Balance for weighing
 Autoclave
 Bunsen burner
 Sterile pipettes
 Sterile pipette tips
 Boiling water bath or microwave oven
 9-cm-diameter bacteriological petri dishes
 Sterile 50-ml glass conical flasks
 Sterile 250-ml glass conical flasks
 Plastic sterile 50-ml tubes
 Plastic sterile 2-ml tubes (cryotubes)
 Shaking incubator at 37°C
 Static incubator at 37°C
 Water bath at 37°C
 Bucket of ice
 Glass spreader (made by bending a Pasteur
   pipette in a Bunsen flame at an angle of
   90°, to give a flat glass surface about 5 cm
   long)
 Beaker with about 50 ml of ethanol
 Refrigerated Sorvall centrifuge at 4°C with
   rotor SS34 (or equivalent)

PROTOCOL A

1. Prepare agar plates as follows:
     (i) Weigh 2.2 g bacto-agar into a 500-ml bottle.
    (ii) Add 200 ml of L-broth.
   (iii) Autoclave the mixture.
    (iv) Boil in a water bath or heat in a microwave oven
         until liquid.
     (v) Cool to 50°C and add antibiotic solution (0.2 ml
         of ampicillin or tetracycline stock solution).
    (vi) Pour 30 ml onto each of six bacteriological petri
         dishes (9-cm size).
   (vii) Flame the surface of the agar in the petri dishes
         with a cool Bunsen flame to remove any air
         bubbles.
  (viii) Leave at room temperature for 24 hours to set
         and dry out.
2. Prepare stock culture of bacteria as follows:
     (i) Place 5 ml of L-broth in a sterile 50-ml glass
         conical flask.

  (ii) Add about a hundred bacterial cells strain HB101.

  (iii) Shake vigorously in an incubator at 37°C overnight.

3. Prepare bacteria for transformation as follows:

  (i) Place 0.5 ml of the overnight bacterial culture in a 250-ml glass conical flask.

  (ii) Add 50 ml of L-broth.

  (iii) Shake vigorously in an incubator at 37°C for 2 hours until the OD_{600} is about 0.5.

  (iv) Transfer to a plastic sterile 50-ml tube.

  (v) Pellet the bacteria in a Sorvall centrifuge (rotor SS34) at 4°C at 10 000 rpm (12 000 g) for 5 minutes.

  (vi) Resuspend bacteria in 50 ml of 80 mM calcium chloride (made by using 2 ml 2 M calcium chloride and 28 ml water).

  (vii) Leave on ice for 15 minutes.

  (viii) Pellet bacteria by centrifugation at 4°C as at step (v).

  (ix) Resuspend bacteria in 5 ml of 80 mM calcium chloride (made by using 0.2 ml 2 M calcium chloride and 4.8 ml water).

  (x) Keep on ice and use within 4 hours.

4. Transform bacteria with plasmid as follows:

  (i) Place 200 μl of treated bacteria in a sterile 2-ml tube.

  (ii) Add 1 μg of plasmid made up to 100 μl with TEN buffer.

  (iii) Leave on ice for 30 minutes.

  (iv) Place in a water bath at 37°C for 5 minutes.

  (v) Add 300 μl of L-broth.

  (vi) Leave in a water bath at 37°C for another 30 minutes (for tetracycline resistance) or another 1 hour (for ampicillin resistance).

5. Plate transformed bacteria onto agar plates containing antibiotic:

  (i) Prepare agar plates with different numbers of bacteria: e.g. place 5 μl onto one dish, 50 μl onto another and 500 μl onto another.

  (ii) Spread bacteria over each dish evenly using a sterile glass spreader. The spreader is sterilised by dipping it in a beaker of ethanol and lighting in a Bunsen flame.

  (iii) Incubate the petri dishes upside down in an incubator at 37°C overnight until bacterial colonies are visible by eye.

B. Growth of bacteria and amplification of plasmid

A single bacterial colony is picked off an agar dish and is amplified in suspension, firstly as a small 5-ml culture and then to much larger volumes.

Solutions and chemicals required:
L-broth (made as in protocol A)
Appropriate antibiotic
  – ampicillin 50 mg/ml in water (filter-
    sterilised through 0.2-μm membrane)
  – tetracycline 15 mg/ml in methanol (filter-
    sterilised through 0.2-μm filter)
Glycerol (autoclaved)

Equipment recommended:
Platinum wire loop on a handle
Bunsen burner
Sterile 50-ml glass conical flasks
Sterile 1-litre glass conical flasks
Shaking incubator at 37°C

PROTOCOL B

1. Place 5 ml of L-broth in a sterile 50-ml glass conical flask.
2. Add 5 μl of stock ampicillin or tetracycline solution.
3. Pick a single bacterial colony off an agar dish with a sterile platinum loop and add it to the L-broth. (The wire loop is sterilised by heating in a Bunsen flame until it is red hot; it must be cooled in air before touching the bacteria.)
4. Shake vigorously in an incubator at 37°C overnight.
5. Next morning, place 400 ml of L-broth in a sterile 1-litre glass conical flask.
6. Add 1.0 ml of stock ampicillin or tetracycline solution.
7. Add 4 ml of the bacterial culture.
8. Shake vigorously in an incubator at 37°C overnight.
9. Next morning, store a 2-ml aliquot of the bacterial cells mixed with 2 ml of glycerol at −20°C as a stock solution for any future use. Such cells can be removed from the freezer and a small (0.1 ml) aliquot amplified whenever needed by returning only to step 5 of protocol B.

C. Large-scale preparation of plasmid DNA

The bacteria are harvested by centrifugation and then lysed in buffer solution containing lysozyme, to break open the bacterial cell walls, and the detergent SDS, to separate out all the cell contents. Addition of potassium acetate results in the formation of a heavy white precipitate of bacterial genomic DNA entangled in the remnants of the bacterial cells. This precipitate is spun out, leaving the plasmid DNA still in solution. Addition of isopropanol to the supernatant then precipitates out the plasmid DNA, which is pelleted by centrifugation. Unfortunately, these precipitations are not absolute and the plasmid DNA has to be purified from cell RNA and remnants of single-stranded bacterial DNA by equilibrium-density-gradient centrifugation in caesium chloride in the presence of saturating quantities of ethidium bromide. Under these conditions, plasmid DNA bands at a higher density than bacterial cell RNA and single-stranded DNA remnants. Usually there are two bands of plasmid DNA about half-way down the gradient. The lower band is the supercoiled plasmid DNA. The upper band is the open circular form of plasmid DNA and should be present in only low amounts in a good plasmid preparation. The lower supercoiled band of plasmid DNA is isolated, ethidium bromide is removed and the DNA is precipitated by addition of ethanol. The plasmid DNA can finally be spun down as a pellet and redissolved in buffer for use.

Quantities given in the following recipes are for preparation of plasmid from a 400-ml bacterial culture. If greater quantities are required the amounts given must be scaled up appropriately.

Equipment recommended:
  Balance for weighing
  Refrigerated Sorvall centrifuge at 4°C with
    rotor GSA (or equivalent)
  250-ml plastic bottles for centrifuge
  Bucket of ice
  100-ml measuring cylinder
  Glass funnel with muslin filter
  10-ml polycarbonate centrifuge tube – Oak
    Ridge type
  Beckman ultracentrifuge with rotor 50Ti (or
    equivalent)
  30-ml Corex glass tubes (autoclaved)
  Ultraviolet light
  Glass Pasteur pipettes (autoclaved)
  −20°C freezer
  Spectrophotometer

Solutions and chemicals required:

Lysozyme (Sigma)

Solution I made as follows:

  4.5 g D-glucose

  12.5 ml 1 M Tris–HCl pH 8.0

  10 ml 0.5 M EDTA

  Make up to 500 ml with distilled water

  Autoclave

Solution II made as follows:

  20 ml 1 M NaOH (autoclaved)

  10 ml 10% w/v SDS (Serva) in water (filter-sterilised 0.2-μm membrane)

  70 ml distilled water (autoclaved)

  This solution should be made freshly as required and in any case not stored for more than one week.

Solution III made as follows:

  147.2 g potassium acetate

  450 ml distilled water

  Adjust pH to 5.0 with glacial acetic acid

  Autoclave

10 × TE buffer made as follows:

  40 ml 1 M Tris–HCl pH 8.0

  8 ml 0.5 M EDTA

  Make volume up to 400 ml with distilled water

  Autoclave

Isopropanol (propan-2-ol)

Ethidium bromide solution 5 mg/ml in water

Caesium chloride

Isobutanol

100% ethanol

70% v/v ethanol in water

Distilled water (autoclaved)

PROTOCOL C

1. Pour the bacterial culture into 250-ml plastic bottles for the centrifuge.
2. Harvest the bacteria by centrifugation at 8000 rpm (10 400 g) for 5 minutes in the Sorvall centrifuge using GSA rotor at 4°C.
3. Pour off the supernatant and drain the pellet.
4. Resuspend the pellet from a 400-ml culture in 20 ml of solution I in a 250-ml plastic centrifuge bottle.
5. Add solid lysozyme to a final concentration of 5 mg/ml and mix to dissolve.

6. Keep at room temperature for 10 minutes.
7. Add 40 ml of solution II and mix well. The solution should turn very viscous at this stage.
8. Keep on ice for 5 minutes.
9. Add 20 ml of solution III and mix well. A heavy white precipitate should form.
10. Keep on ice for 15 minutes.
11. Spin down the precipitate at 8000 rpm (10 400 g) for 5 minutes in the Sorvall centrifuge using GSA rotor at 4°C.
12. Take the supernatant. Pour it into a measuring cylinder through a muslin filter in a glass funnel to remove any lumps of precipitate. Measure the volume and transfer to a clean 250-ml plastic centrifuge bottle.
13. Add 0.6 volumes of isopropanol. A fine precipitate of plasmid DNA should now form.
14. Spin down the precipitate at 8000 rpm (10 400 g) for 10 minutes in the Sorvall centrifuge using GSA rotor at 4°C.
15. Discard the supernatant by pouring off. Drain and dry the pellet, either for a few minutes in air or in a vacuum desiccator if available.
16. Redissolve the precipitate in 5.4 ml of 10 × TE buffer and transfer to a Corex glass tube.
17. Add 0.6 ml of ethidium bromide (5 mg/ml) solution.
18. Add 6.1 mg of solid caesium chloride.
19. Transfer to a 10-ml polycarbonate tube (Oak Ridge type).
20. At this stage, all tubes should be balanced to within 0.1 g ready for centrifugation and tubes should be capped. Tubes should also be filled to within 1 ml of full capacity or else they can collapse during centrifugation − this can be done by overlaying with liquid paraffin if necessary.
21. Centrifuge at 35 000 rpm for at least 40 hours (or for convenience over the weekend) at 15°C in a Beckman ultracentrifuge in a 50Ti rotor (111 000−51 300 g).
22. Remove tubes from the centrifuge and view in ultraviolet light. The plasmid DNA should be clearly visible as a single red band of supercoiled form in the middle of the gradient, although a second minor band of open circular plasmid is sometimes also obtained just above. The heavy red pellet at the base of the tube is of bacterial RNA and single-stranded DNA.
23. Collect the band of supercoiled plasmid DNA using a Pasteur pipette and transfer to a 30-ml Corex glass tube.

24. Add an equal volume of isobutanol and vortex. Allow to stand for one minute to separate the phases. Discard the upper (pink) isobutanol phase.

25. Repeat step 24 twice more until all ethidium bromide has been removed from the plasmid DNA solution.

26. Dilute the plasmid DNA solution 3- to 4-fold with distilled water and transfer to a clean Corex glass tube. Measure the volume and add 2 volumes of 100% ethanol.

27. Keep at $-20°C$ for 2 hours. The mixture must not be stored for any prolonged period or the caesium chloride will come out of solution.

28. Spin down the DNA precipitate at 10 000 rpm (12 000 g) at 4°C for 20 minutes in the Sorvall centrifuge (rotor SS34) or equivalent.

29. Wash the precipitate once with 70% ethanol and repeat step 28.

30. Drain and dry the DNA pellet for a few minutes at room temperature and redissolve it in 0.5 ml of distilled water.

31. The concentration of the DNA can be measured using optical density:

    Take 10 μl of plasmid DNA solution.
    Add 490 μl of distilled water.
    Read optical density at 260 nm (OD_{260}).
    A solution of DNA in water of 1 mg/ml gives an OD_{260} of 20.
    Thus $\mu l/\mu l$ in DNA solution $= 1/20 \times OD_{260} \times 50$.

32. Store at $-20°C$.

D.   Isolation of cloned gene sequences

The final steps involve purification of the cloned gene sequences from the plasmid. The plasmid DNA is cut with the appropriate restriction enzyme(s) used for the original gene cloning. This releases the cloned gene from the plasmid and the two DNAs can be separated by agarose gel electrophoresis.

Nuclease contamination must be avoided in procedures 1 and 3. For these protocols, gloves must be worn throughout; pipette tips, glassware and solutions should all be autoclaved before use. These precautions are not necessary during the running of the agarose gel (procedure 2).

Solutions and chemicals required:

10 × restriction enzyme (10 × RE) buffer made as follows:

    2 ml 1 M Tris−HCl pH 7.4 (autoclaved)

    1.6 ml distilled water (autoclaved)

    0.4 ml 1 M magnesium chloride (autoclaved)

    28 μl stock β-mercaptoethanol

    This can be stored at −20°C for up to one month

Restriction enzyme preparation (Boehringer or Biolabs)

5 M sodium chloride (autoclaved)

Distilled water (autoclaved)

5 × TBE buffer made as follows:

    106 g Tris base

    55 g Boric acid

    9.3 g dipotassium EDTA

    Dissolve and make up to 2 litres with distilled water

1 × TBE buffer made by dilution of 5 × TBE buffer

Ethidium bromide (BDH) 10 mg/ml in water (store in the dark)

Agarose powder (SEAKEM-ME) (FMC Corp.)

DNA sample buffer (DNA SB) made as follows:

    330 μl glycerol

    75 μl 5 × TBE buffer

    250 μl 0.5 M EDTA (pH 7.0)

    Mix thoroughly. Then

    Add 27 μl of 10% w/v SDS (Serva) in water (if this is added earlier it precipitates out)

    Add bromophenol-blue powder (spatula tip, approx. 100 μg)

Phenol, redistilled and water-saturated (Rathburn Chemicals)

Chloroform

3 M sodium acetate pH 5.2

100% ethanol

70% v/v ethanol in water

Equipment recommended:
Microcentrifuge tubes 1.5 ml volume (auto-
  claved) (Treff Lab)
Water bath at 37°C
Microwave oven or boiling water bath
Agarose gel electrophoresis equipment
  (commercially available or home-made)
Filter-paper wicks
Electrophoresis power supply
Ultraviolet light
Scalpel
Microcentrifuge (Eppendorf or M.S.E.)
Freezer at −20°C

PROTOCOL D

1. The plasmid DNA is cut with the appropriate restriction enzyme(s) used for the original gene cloning, as
follows:
  (i) Place 20 μg of plasmid DNA in a 1.5-ml
    microcentrifuge tube.
  (ii) Add 5 μl of 10 × RE buffer.
  (iii) Add the required amount of enzyme (up to 5 μl
    volume only). The amount of enzyme required
    is calculated in units (1 unit of enzyme cuts 1 μg
    of DNA in 1 hour at 37°C in a volume of 50 μl).
  (iv) Add 5 M NaCl as needed by the enzyme. There
    are three groups of enzyme:
      Low-salt enzymes   − add no 5 M NaCl
      Medium-salt enzymes − add 0.5 μl of 5 M
                              NaCl
      High-salt enzymes   − add 1.0 μl of 5 M
                              NaCl
    The salt requirements are specified by the
    supplier.
  (v) Add distilled water to make the volume up to
    50 μl.
  (vi) Incubate reaction mixture for 1 hour minimum
    or if convenient overnight. Most enzymes work
    best at 37°C but some require other tempera-
    tures. This will be specified by the supplier.

2. The cloned gene is separated from plasmid sequences on a 0.8% agarose gel, as follows:

   (i) Prepare mould for the gel. Many systems are available commercially for pouring gels but a simple home-made system is illustrated in Fig. 6.3.

   A perspex plate is levelled using a spirit level. A glass plate (14 cm × 19 cm) is used as the mould. Its edges are sealed with PVC electrical tape (or radioactive/biohazard-labelled tape) so that it forms a wall 1 cm high all around the plate. This is placed on the level perspex plate. A simple perspex comb is assembled at one end of the plate so that there is about 1 mm of space between the base of the teeth and the glass plate.

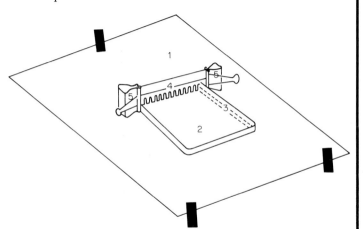

Fig. 6.3 Mould for pouring agarose gels.
1 = perspex plate levelled with three screw feet;
2 = glass plate 14 cm × 19 cm; 3 = PVC tape;
4 = perspex comb to mould wells in the gel;
5 = clamp to hold comb in position

   (ii) Weigh 0.8 g of agarose into a 200 ml conical flask.

   Add 100 ml of 1 × TBE buffer.

   Add 100 µl of ethidium bromide solution (10 mg/ml in water).

   Heat in microwave oven or boiling water bath until agarose dissolves.

   (iii) Pour agarose onto gel mould and allow to set for 1–2 hours.

   NB The recipe given here provides a 0.8% agarose gel, which is useful for most purposes.

To separate very large or very small DNA fragments, the concentration of agarose can be varied.

(For further details see Maniatis *et al.*, 1982).

(iv) The gel tank is prepared and connected to an electrophoresis power supply. Many systems are available commercially but a simple home-made system is illustrated in Fig. 6.4.

(v) Place 1600 ml of 1 × TBE buffer into the gel tank and add 2 drops of ethidium bromide solution.

(vi) Put the gel into the tank. The surface of the gel should be just submerged (by about 1 cm depth of buffer).

(vii) To the 50 μl enzyme reaction mixture, add 20 μl of DNA SB.

(viii) Load the DNA into wells of the gel.

(ix) Load also some polynucleotides of known size in an adjacent well to provide molecular weight markers. Common markers used are phage λ DNA cut with the restriction enzyme HindIII (Biolabs). (See Table 3.1.)

(x) Run the gel for 2—4 hours at 120 V. All samples are run from the negative to the positive electrode.

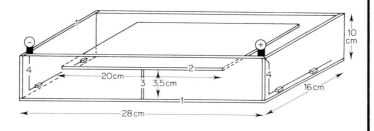

Fig. 6.4 Agarose gel electrophoresis system. A perspex box (1) made of 0.6 cm thick perspex, with a lid, has an internal perspex shelf (2) moulded to the sides of the box. The box is divided into two compartments with a perspex divider (3). Two electrodes are made of platinum wire (4), one in each chamber

3. The cloned gene is eluted from the gel as follows:
    (i) View gel under ultraviolet light to locate the DNA bands. Cut, with a scalpel blade, a small trough right through the gel immediately in front of the band to be eluted (on the positive electrode side of the band). The trough should be the length of the DNA band and about 3–5 mm wide.
   (ii) Replace the gel onto a clean dry glass plate. Trickle some liquid agarose (0.8% in 1 × TBE buffer as at step 2(ii)) around the base of the trough to seal it. Allow to set for 5 minutes.
  (iii) Fill the well with 1 × TBE buffer.
   (iv) Place the gel back into the gel tank but with only 200 ml of 1 × TBE buffer, so that the gel is not covered with any buffer.
    (v) Soak two pieces of filter paper in 1 × TBE buffer and assemble them as wicks from the gel to the buffer tank. This assembly is illustrated in Fig. 6.5.
   (vi) Subject to electrophoresis at 500 V for 30-second bursts. View the gel after each burst to determine exactly where the DNA band is. When the DNA has moved into the buffer in the well, transfer it to a 1.5 ml microcentrifuge tube. If the DNA moves too far at any stage, the current can be reversed until the DNA is back in the well.

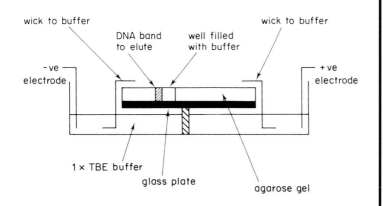

Fig. 6.5 Assembly for elution of DNA bands from agarose gels

4. The eluted DNA is finally purified as follows:

    (i) To the DNA solution in the microcentrifuge tube, add an equal volume of phenol.

    (ii) Mix by vortexing.

    (iii) Break the phases by centrifugation for 5 minutes in a microcentrifuge.

    (iv) Transfer the upper aqueous layer to a clean 1.5-ml microcentrifuge tube.

    (v) Repeat steps 4(i)–4(iv) once.

    (vi) Add an equal volume of chloroform.

    (vii) Mix by vortexing.

    (viii) Break the phases by centrifugation for 5 minutes in a microcentrifuge.

    (ix) Transfer the upper aqueous layer to a clean 1.5-ml microcentrifuge tube.

    (x) Add 1/10 × volume of 3 M sodium acetate pH 5.2 and mix.

    (xi) Add 2 volumes of 100% ethanol.

    (xii) Store at $-20°C$ overnight (about 18 hours). If the DNA is required urgently, the tube can be placed in dry ice and the DNA is then recoverable after 30 minutes.

    (xiii) Spin down the DNA precipitate in a microcentrifuge for 15 minutes.

    (xiv) Pour off the ethanol.

    (xv) Wash the DNA pellet in 70% ethanol.

    (xvi) Spin down the DNA precipitate in a microcentrifuge for 10 minutes.

    (xvii) Pour off the ethanol. Drain and dry the DNA pellet for a few minutes.

    (xviii) Redissolve the DNA in 20 μl of sterile distilled water.

    (xix) Store at $-20°C$ and use in the nick translation reaction as required (Chapter 3, protocol E; Chapter 5, protocol C).

6.4 REFERENCES

Cohen, S.N., Chang, A.C.Y. and Hsu, L. (1973) *Proc. Natl. Acad. Sci. USA* **69:** 2110–2114.

Mandel, M. and Higa, A. (1970) *J. Mol. Biol.* **53:** 159–162.

Maniatis, T., Fritsch, E.F. and Sambrook, J. (1982) In: *Molecular Cloning: A Laboratory Manual.* Cold Spring Harbor Laboratory, Cold Spring Harbor, NY.

Radloff, R., Bauer, W. and Vinograd, J. (1967) *Proc. Natl. Acad. Sci. USA* **57:** 1514–1521.

Appendix A

Abbreviations

A	adenine
ADP	adenosine diphosphate
ATP	adenosine triphosphate
BSA	bovine serum albumin
C	cytosine
CaCl$_2$	calcium chloride
cm	centimetre
cpm	counts per minute
CsCl	caesium chloride
cDNA	complementary or copy DNA
dATP	deoxyadenosine triphosphate
dCTP	deoxycytidine triphosphate
dGTP	deoxyguanosine triphosphate
DNA	deoxyribonucleic acid
DNase	deoxyribonuclease
DS	double-stranded
dTTP	deoxythymidine triphosphate
EDTA	Ethylenediamine tetraacetic acid
E. coli	*Escherichia coli*
G	guanine
g	gram
HCl	hydrochloric acid
kb	kilobase (number of bases in thousands)
kg	kilogram

LiCl	lithium chloride
M	molar
mA	milliamp
mg	milligram
μg	microgram
Mg^{++}	magnesium ion
$MgCl_2$	magnesium chloride
min	minute
ml	millilitre
μl	microlitre
MOPS	3-(N-morpholino)propanesulphonic acid
mm	millimetre
mM	millimolar
μM	micromolar
μm	micrometre (10^{-6} metre)
mRNA	messenger ribonucleic acid
NaCl	sodium chloride
NaOH	sodium hydroxide
nm	nanometre
nM	nanomolar
OD	optical density
OD_{260}	optical density at 260 nm wavelength
OD_{600}	optical density at 600 nm wavelength
OH	hydroxyl group
oligo (dT)	oligodeoxythymidylic acid
^{32}P	phosphorus-32
PBS	phosphate-buffered saline
pg	picogram
poly(A)	polyadenylic acid
poly(A)$^+$	Polyadenylated mRNA
RNA	ribonucleic acid
RNase	ribonuclease
rpm	revolutions per minute
rRNA	ribosomal RNA
S	Svedberg unit (sedimentation coefficient of 1×10^{-13} sec)
SDS	sodium dodecyl sulphate
SS	single-stranded
T	thymine
Tris	Tris(hydroxymethyl)aminomethane
Tris–HCl	Tris(hydroxymethyl)aminomethane hydrochloride
t-RNA	transfer ribonucleic acid
U	uracil
uv	ultraviolet (light)
V	volt
v/v	volume for volume (as in %v/v, which indicates the volume in ml in a 100 ml total volume)
w/v	weight for volume (as in %w/v, which indicates the number of grams in a 100 ml total volume)

Appendix B

Source of reagents and equipment

Where specific manufacturers are recommended within the text, addresses are given here. This implies only that the recommended source has proved reliable in recent years in my experience.

Amersham	Amersham International plc, Amersham, England.
Beckman	Beckman Instruments Inc., 1117 California Avenue, Palo Alto, CA 94304, USA. (UK supplier: Beckman RIIC Ltd., Progress Road, Sands Ind. Estate, High Wycombe, Bucks., England.)
Biolabs	New England Biolabs Inc., 32 Tozer Road, Beverly, MA 01915, USA. (UK supplier: CP Laboratories Ltd., PO Box 22, Bishops Stortford, Herts., England.)
Boehringer	Boehringer Mannheim GmbH, Biochemica, D6800 Mannheim 31, West Germany. (UK supplier: BCL Boehringer, Mannheim House, Bell Lane, Lewes, East Sussex, England.)
Collaborative Research	Collaborative Research, 128 Spring Street, Lexington, MA 02173, USA. (UK supplier: Stratech Scientific, 50 Newington Green, London N16, England.)
Dupont (Sorvall)	Dupont (UK), Wedgewood Way, Stevenage, Herts., England.
Eppendorf	Eppendorf Gerätebau, Nether & Hinz GmbH, Postf 65 06 70, 2000 Hamburg 65, West Germany. (UK supplier: Anderman & Co. Ltd., 145 London Road, Kingston-upon-Thames, Surrey, England.)

Falcon	Falcon is a trademark of Becton, Dickinson & Co. (UK). Between Towns Road, Cowley, Oxford, England.
Fluka	Fluka AG, CH-9470 Buchs, Switzerland. (UK supplier: Fluka, Peak Dale Road, Glossop, Derbyshire, England.)
FMC Corp.	FMC Bio Products, 5 Maple Street, Rockland, ME 04841, USA. (UK supplier: ICN Biomedicals Ltd., Free Press House, Castle Street, High Wycombe, Bucks., England.)
Kodak	Eastman Kodak Company, Rochester, NY 14650, USA. (UK supplier: Kodak Ltd., PO Box 33, Swallowdale Lane, Hemel Hempstead, Herts., England.)
M.S.E.	M.S.E. Scientific Instruments, Sussex Manor Park, Gatwick Road, Crawley, Sussex, England.
Pharmacia	Pharmacia, S-75182 Uppsala, Sweden. (UK supplier: Pharmacia Ltd., Pharmacia House, Midsummer Boulevard, Central Milton Keynes, Bucks., England.)
Rathburn Chemicals	Rathburn Chemicals Ltd., Walkerburn, Peeblesshire, Scotland.
Sartorius	Sartorius GmbH, PF 3243, D-3400 Göttingen, West Germany. (UK supplier: Sartorius, 18 Avenue Road, Belmont, Surrey, England.)
Serva	Serva Feinbiochemica, Heidelberg, West Germany. (UK supplier: Uniscience, 12–14 St Ann's Crescent, London, England.)
Sigma	Sigma Chemical Company Ltd., Fancy Road, Poole, Dorset, England.
Treff Lab	Treff AG, CH-9113 Degersheim, Switzerland. (UK supplier: Scotlab, Unit 15 Earn Avenue, Righead Industrial Estate, Bellshill, ML4 3JQ, England.)
Whatman	Whatman International Ltd., Maidstone, England.
Worthington	Worthington Biochemical Corp., Freehold, New Jersey 07728, USA. (UK supplier: Cambridge BioScience, 42 Devonshire Road, Cambridge, England.)

Appendix C

Recipes for routine solutions

2 M Calcium chloride

Dissolve 29.4 g of $CaCl_2 \cdot 2H_2O$ in distilled water and make up to a final volume of 100 ml. Filter-sterilise through a 0.2 μm membrane.

0.5 M Dithiothreitol (DTT)

Dissolve 1.55 g of DTT in 20 ml of distilled water and 67 μl of 3 M sodium acetate pH 5.2. Sterilise by filtration through a 0.2 μm membrane and store in 0.5 -ml aliquots at $-20°C$.

0.5 M EDTA (pH 7.0, 7.5 or 8.0)

Dissolve 18.6 g of the disodium salt of EDTA in 80 ml of distilled water. Adjust the pH as desired with either very strong NaOH solution or NaOH pellets. Adjust the final volume to 100 ml with distilled water. Sterilise by autoclaving.

This compound will not dissolve completely until the pH is adjusted with NaOH to near pH 7.0.

1 M Magnesium chloride

Dissolve 20.3 g of $MgCl_2 \cdot 6H_2O$ in distilled water and make up to a final volume of 100 ml. Sterilise by autoclaving.

0.5 M Magnesium chloride

Dissolve 10.2 g of $MgCl_2 \cdot 6H_2O$ in distilled water and make up to a final volume of 100 ml. Sterilise by autoclaving.

Phosphate-buffered saline (PBS)

Dissolve 10 g NaCl + 0.25 g KCl + 1.43 g Na_2HPO_4 + 0.25 g KH_2PO_4 in distilled water and make up to a final volume of 1 litre. Check the pH is 7.2. Sterilise by autoclaving.

3 M Potassium acetate (pH 5.0)

Dissolve 29.4 g of potassium acetate (anhydrous) in 80 ml distilled water. Adjust the pH to 5.0 with glacial acetic acid and make the final volume up to 100 ml with distilled water. Sterilise by autoclaving.

3 M Sodium acetate (pH 5.2)

Dissolve 24.6 g of sodium acetate (anhydrous) in 80 ml distilled water. Adjust the pH to 5.2 with glacial acetic acid and make the final volume up to 100 ml with distilled water. Sterilise by autoclaving.

5 M Sodium chloride

Dissolve 29.0 g of sodium chloride in distilled water and make up to a final volume of 100 ml. Sterilise by autoclaving.

50 mM Sodium citrate (pH 7.0)

Dissolve 1.47 g of trisodium citrate dihydrate in 80 ml distilled water. Adjust the pH to 7.0 and make the final volume up to 100 ml. Sterilise by autoclaving.

10% SDS

Dissolve 10 g of sodium dodecyl sulphate (SDS) (Serva, high purity grade) in distilled water and make up to a final volume of 100 ml. This solution cannot be autoclaved. It is, however, desirable to filter-sterilise through a 0.2 μm membrane.

5 M Sodium hydroxide

Dissolve 20 g of NaOH in distilled water and make up to a final volume of 100 ml. Sterilise by autoclaving.

1 M Sodium hydroxide

Dissolve 4 g of NaOH in distilled water and make up to a final volume of 100 ml. Sterilise by autoclaving.

0.1 M Sodium hydroxide

Mix 10 ml of 1 M NaOH with 90 ml of distilled water. Sterilise by autoclaving.

100 mM Spermidine

Dissolve 255 mg of spermidine trihydrochloride (Sigma) in distilled water and make up to a final volume of 10 ml. Sterilise by filtration through a 0.2 μm membrane and store in aliquots at $-20°C$.

1 M Tris−HCl (pH 7.4, 8.0 or 9.5)

Dissolve 12.1 g of Tris base in 80 ml of distilled water. Adjust the pH as desired using concentrated HCl and make the final volume up to 100 ml with distilled water. Sterilise by autoclaving.

10 mM Tris−HCl (pH 7.5 or 8.0)

Dissolve 0.12 g of Tris base in 80 ml of distilled water. Adjust the pH as desired using concentrated HCl and make the final volume up to 100 ml with distilled water. Sterilise by autoclaving.

0.5 M Tris−HCl (pH 8.0) containing 1 mM EDTA

Dissolve 6.05 g of Tris base and 20.3 mg of disodium EDTA in 80 ml of distilled water. Adjust the pH to 8.0 using concentrated HCl and make the final volume up to 100 ml with distilled water. Sterilise by autoclaving.

10 mM Tris−HCl (pH 8.0) containing 1 mM EDTA

Dissolve 0.12 g of Tris base and 20.3 mg of disodium EDTA in 80 ml of distilled water. Adjust the pH to 8.0 using concentrated HCl and make the final volume up to 100 ml with distilled water. Sterilise by autoclaving.

Index